Geometry

**LARSON
BOSWELL
STIFF**

Applying • Reasoning • Measuring

Chapter 1 Resource Book

The Resource Book contains the wide variety of blackline masters available for Chapter 1. The blacklines are organized by lesson. Included are support materials for the teacher as well as practice, activities, applications, and assessment resources.

CURRICULUM
MSSU

McDougal Littell
A HOUGHTON MIFFLIN COMPANY
Evanston, Illinois • Boston • Dallas

Contributing Authors

The authors wish to thank the following individuals for their contributions to the Chapter 1 Resource Book.

Eric J. Amendola
Karen Collins
Michael Downey
Patrick M. Kelly
Edward H. Kuhar
Lynn Lafferty
Dr. Frank Marzano
Wayne Nirode
Dr. Charles Redmond
Paul Ruland

Copyright © 2001 McDougal Littell Inc.
All rights reserved.

Permission is hereby granted to teachers to reprint or photocopy in classroom quantities the pages or sheets in this work that carry a McDougal Littell copyright notice. These pages are designed to be reproduced by teachers for use in their classes with accompanying McDougal Littell material, provided each copy made shows the copyright notice. Such copies may not be sold and further distribution is expressly prohibited. Except as authorized above, prior written permission must be obtained from McDougal Littell Inc. to reproduce or transmit this work or portions thereof in any other form or by any other electronic or mechanical means, including any information storage or retrieval system, unless expressly permitted by federal copyright laws. Address inquiries to Manager, Rights and Permissions, McDougal Littell Inc., P.O. Box 1667, Evanston, IL 60204.

ISBN: 0-618-02064-0

6789-VEI- 04 03

Contents

1 Basics of Geometry

Chapter Support	1–8
1.1 Patterns and Inductive Reasoning	9–21
1.2 Points, Lines, and Planes	22–33
1.3 Segments and Their Measures	34–48
1.4 Angles and Their Measures	49–60
1.5 Segment and Angle Bisectors	61–77
1.6 Angle Pair Relationships	78–91
1.7 Introduction to Perimeter, Circumference, and Area	92–108
Review and Assess	109–122
Resource Book Answers	A1–A14

Contents

CHAPTER SUPPORT MATERIALS

Tips for New Teachers	p. 1	Prerequisite Skills Review	p. 5
Parent Guide for Student Success	p. 3	Strategies for Reading Mathematics	p. 7

LESSON MATERIALS

	1.1	1.2	1.3	1.4	1.5	1.6	1.7
Lesson Plans (Reg. & Block)	p. 9	p. 22	p. 34	p. 49	p. 61	p. 78	p. 92
Warm-Ups & Daily Quiz	p. 11	p. 24	p. 36	p. 51	p. 63	p. 80	p. 94
Alternative Lesson Openers	p. 12	p. 25	p. 37	p. 52	p. 64	p. 81	p. 95
Tech. Activities & Keystrokes					p. 65	p. 82	p. 96
Practice Level A	p. 14	p. 26	p. 38	p. 53	p. 69	p. 84	p. 100
Practice Level B	p. 15	p. 27	p. 39	p. 54	p. 70	p. 85	p. 101
Practice Level C	p. 16	p. 28	p. 40	p. 55	p. 71	p. 86	p. 102
Reteaching and Practice	p. 17	p. 29	p. 41	p. 56	p. 72	p. 87	p. 103
Catch-Up for Absent Students	p. 19	p. 31	p. 43	p. 58	p. 74	p. 89	p. 105
Coop. Learning Activities				p. 44			p. 106
Interdisciplinary Applications		p. 32		p. 59		p. 90	
Real-Life Applications	p. 20		p. 45		p. 75		p. 107
Math and History Applications			p. 46				
Challenge: Skills and Appl.	p. 21	p. 33	p. 47	p. 60	p. 76	p. 91	p. 108

REVIEW AND ASSESSMENT MATERIALS

Quizzes	p. 48, p. 77	Alternative Assessment & Math Journal	p. 117
Chapter Review Games and Activities	p. 109	Project with Rubric	p. 119
Chapter Test (3 Levels)	p. 110	Cumulative Review	p. 121
SAT/ACT Chapter Test	p. 116	Resource Book Answers	p. A1

Contents

Descriptions of Resources

This Chapter Resource Book is organized by lessons within the chapter in order to make your planning easier. The following materials are provided:

Tips for New Teachers These teaching notes provide both new and experienced teachers with useful teaching tips for each lesson, including tips about common errors and inclusion.

Parent Guide for Student Success This guide helps parents contribute to student success by providing an overview of the chapter along with questions and activities for parents and students to work on together.

Prerequisite Skills Review Worked-out examples are provided to review the prerequisite skills highlighted on the Study Guide page at the beginning of the chapter. Additional practice is included with each worked-out example.

Strategies for Reading Mathematics The first page teaches reading strategies to be applied to the current chapter and to later chapters. The second page is a visual glossary of key vocabulary.

Lesson Plans and Lesson Plans for Block Scheduling This planning template helps teachers select the materials they will use to teach each lesson from among the variety of materials available for the lesson. The block-scheduling version provides additional information about pacing.

Warm-Up Exercises and Daily Homework Quiz The warm-ups cover prerequisite skills that help prepare students for a given lesson. The quiz assesses students on the content of the previous lesson. (Transparencies also available)

Activity Support Masters These blackline masters make it easier for students to record their work on selected activities in the Student Edition.

Alternative Lesson Openers An engaging alternative for starting each lesson is provided from among these four types: *Application, Activity, Geometry Software,* or *Visual Approach.* (Color transparencies also available)

Technology Activities with Keystrokes Keystrokes for Geometry software and calculators are provided for each Technology Activity in the Student Edition, along with alternative Technology Activities to begin selected lessons.

Practice A, B, and C These exercises offer additional practice for the material in each lesson, including application problems. There are three levels of practice for each lesson: A (basic), B (average), and C (advanced).

Contents

Reteaching with Practice These two pages provide additional instruction, worked-out examples, and practice exercises covering the key concepts and vocabulary in each lesson.

Quick Catch-Up for Absent Students This handy form makes it easy for teachers to let students who have been absent know what to do for homework and which activities or examples were covered in class.

Cooperative Learning Activities These enrichment activities apply the math taught in the lesson in an interesting way that lends itself to group work.

Interdisciplinary Applications/Real-Life Applications Students apply the mathematics covered in each lesson to solve an interesting interdisciplinary or real-life problem.

Math and History Applications This worksheet expands upon the Math and History feature in the Student Edition.

Challenge: Skills and Applications Teachers can use these exercises to enrich or extend each lesson.

Quizzes The quizzes can be used to assess student progress on two or three lessons.

Chapter Review Games and Activities This worksheet offers fun practice at the end of the chapter and provides an alternative way to review the chapter content in preparation for the Chapter Test.

Chapter Tests A, B, and C These are tests that cover the most important skills taught in the chapter. There are three levels of test: A (basic), B (average), and C (advanced).

SAT/ACT Chapter Test This test also covers the most important skills taught in the chapter, but questions are in multiple-choice and quantitative-comparison format. (See *Alternative Assessment* for multi-step problems.)

Alternative Assessment with Rubrics and Math Journal A journal exercise has students write about the mathematics in the chapter. A multi-step problem has students apply a variety of skills from the chapter and explain their reasoning. Solutions and a 4-point rubric are included.

Project with Rubric The project allows students to delve more deeply into a problem that applies the mathematics of the chapter. Teacher's notes and a 4-point rubric are included.

Cumulative Review These practice pages help students maintain skills from the current chapter and preceding chapters.

CHAPTER 1

Tips for New Teachers
For use with Chapter 1

LESSON 1.1

TEACHING TIP While completing Example 3 on page 4, remind students that n is a variable and could represent any number, such as 7, 25, 100, or even 10,500. It is used to represent the general case for this pattern of odd integers. Students can test their conjecture about the sum of the first n odd integers and verify that the pattern continues by using values for n other than those given in the example.

TEACHING TIP Continue to stress that a conjecture can be proven true only when it is true in all cases. Students would not want to try to write and test their conjecture for every whole number n in Example 3. This example leads nicely into the relevance of the counterexample that follows in the next paragraph.

LESSON 1.2

TEACHING TIP Looking at diagrams and their labels can be confusing to students. On page 10 they need to recognize that capital letters are used to identify points, and that a line may be named by any two points on the line or by using a single lower-case script letter. Also explain the difference between the labels of plane M and plane ABC.

LESSON 1.3

TEACHING TIP Make students aware that absolute value is used in the Ruler Postulate because the length of a segment is always expressed as a positive number. In Example 1 on page 17, the length of AB could have been shown as $3 - 5.5$, but a length of -2.5 would not make sense. By using the absolute value of the differences, the result will always be positive regardless of the order in which the coordinates are subtracted.

INCLUSION Subscript notation can be confusing for students. Remind students that the variables in the distance formula such as x_1 can be read or thought of as "the x-coordinate of the first point." You might want to write the formula labeling each of the variables and then have the class recite the formula using the word description.

TEACHING TIP Have students note that the postulates are numbered and that this is done just for easy reference in the textbook. Students should know the meaning of each postulate and be able to recite or write it. Giving a reason by indicating "Postulate 3" is not appropriate.

LESSON 1.4

COMMON ERROR Students tend to think of $=$ and $\cong$ as the same. Emphasize that numeric measurements of geometric shapes (such as angle measurements, line segment lengths, areas, perimeters, etc.) can be equal, but that "congruent to" is used when geometric figures are compared and found to have the same size and the same shape.

TEACHING TIP Stress that students say, "the measure of angle A is equal to the measure of angle B" when the size of the angles is considered. Caution students about writing the angle symbol carefully. Very often it may look like an inequality symbol and therefore be confusing to the teacher and the students. Some teachers ask students to write the angle symbol and put a curved arc through it so it won't be confused with the inequality symbol for "less than."

TEACHING TIP When classifying angles, some students may ask if it is possible to have angles with negative measures or measures greater than 180°. This question might also arise because your geometry course is taught after a course in Algebra II. Of course, the answer is yes, but explain that the description used here is appropriate for the study of plane geometry. Let students know that descriptions and definitions might be extended or modified as other topics in mathematics are studied.

LESSON 1.5

TEACHING TIP Students need to recognize and use the algebraic form of the Midpoint Formula. They should also understand that the x-coordinate

CHAPTER 1 CONTINUED

Tips for New Teachers
For use with Chapter 1

of the midpoint is the average of the x-coordinates of the endpoints of the segment and likewise for the y-coordinate of the midpoint.

TEACHING TIP In Example 5 on page 37, point out that solving an equation for x is not the same as finding the measures of the angles. Checking the value of the variable by determining each angle measure is always important.

LESSON 1.6

COMMON ERROR Students may think that vertical angles are adjacent when viewing a diagram. It may appear to them that a pair of opposite rays of the vertical angles is a common side because the rays form a line. Be sure they understand the difference between adjacent and vertical angles.

INCLUSION Help students identify types of angles by using many visual examples drawn with a variety of orientations. Students with limited English proficiency can recognize the visual example and associate the description of the angle with it. If they have trouble when looking at angles printed on a page or on their paper, encourage them to turn their book or paper and view the angles from other directions or perspectives.

LESSON 1.7

TEACHING TIP Students may be familiar with problem-solving plans from their study of algebra. The plan on page 53 is a little different in steps 3 and 4. You should be aware that theorems and proofs are addressed in Chapter 2, but indicated here for them to consider as part of a general problem-solving plan for geometry. As you review the plan with students, remind them to give their conclusion in sentence form and to be sure the conclusion is appropriate or reasonable for their problem.

Outside Resources

BOOKS/PERIODICALS

Fuys, David, Dorothy Geddes, and Rosamond Tischler. *The van Hiele Model of Thinking in Geometry among Adolescents.* Third in the JRME monograph series. Reston, VA; NCTM.

Coxford, Arthur F, Jr with Linda Burks, Claudia Giamati, and Joyce Jonik. *Geometry from Multiple Perspectives: Addenda Series, Grades 9–12.* Includes activity sheets and helpful instructional suggestions. Reston, VA; NCTM.

SOFTWARE

Spring Branch Software, Inc. *Tools of Mathematics: Points, Lines, and Planes.* Macintosh. Acton, MA. William K. Bradford Publishing Company.

VIDEOS

Mathematics: Making the Connection. Well-known individuals describe how they connect mathematics to their own lives and professions. Reston, VA; NCTM.

CHAPTER 1

NAME _____ DATE _____

Parent Guide for Student Success

For use with Chapter 1

Chapter Overview One way that you can help your student succeed in Chapter 1 is by discussing the lesson goals in the chart below. When a lesson is completed, ask your student to interpret the lesson goals for you and to explain how the mathematics of the lesson relates to one of the key applications listed in the chart.

Lesson Title	Lesson Goals	Key Applications
1.1: Patterns and Inductive Reasoning	Find and describe patterns. Use inductive reasoning to make real-life conjectures.	• Moon Cycles • Bacteria Growth • Molecular Compounds
1.2: Points, Lines, and Planes	Understand and use basic undefined terms and defined terms of geometry. Sketch the intersections of lines and planes.	• Photographs • Sculpture • Perspective Drawing
1.3: Segments and Their Measures	Use segment postulates. Use the Distance Formula to measure distances.	• Map Reading • Incline Railway • Long-Distance Rates
1.4: Angles and Their Measures	Use angle postulates and classify angles as acute, right, obtuse, or straight.	• Angle of Vision • Geography • Airport Runways
1.5: Segment and Angle Bisectors	Bisect a segment and bisect an angle.	• Kite Design • Strike Zone • Paper Airplanes
1.6: Angle Pair Relationships	Identify vertical angles and linear pairs. Identify complementary and supplementary angles.	• Stair Railings • Bridges • Planting Trees
1.7: Introduction to Perimeter, Circumference, and Area	Find the perimeter and area of common plane figures. Use a general problem-solving plan.	• Flag Design • Millennium Dome • Cranberry Harvest

Study Strategy

> **Learning Vocabulary** is the study strategy featured in Chapter 1 (see page 2). Be sure your student has set up a notebook in which to keep math notes, with a section for definitions of new words. Encourage your student to make sure that vocabulary notes are clear and up-to-date, with appropriate use of diagrams. Your student can use these vocabulary notes to share with you what was done in class and to study for tests and quizzes.

Parent Guide for Student Success

For use with Chapter 1

Key Ideas Your student can demonstrate understanding of key concepts by working through the following exercises with you.

Lesson	Exercise
1.1	Describe a pattern in the sequence of numbers. Predict the next number. 4, 4, 8, 12, 20, 32, . . .
1.2	Give an example of the intersection of two planes in a room in your house.
1.3	Use the Distance Formula to decide whether $\overline{PQ} \cong \overline{QR}$ for $P(2, 3)$, $Q(4, -2)$, and $R(-1, 0)$. Explain.
1.4	Think about the angle between the hands of a clock at 10:00. Is the angle *acute*, *right*, *obtuse*, or *straight*? What about at 7:00?
1.5	Segment PQ has midpoint M. P has coordinates $(-3, 5)$. M has coordinates $(7, -1)$. Find the coordinates of Q.
1.6	Main Street in a certain town is straight. Water Street dead ends into Main Street so the angle on one corner is twice the angle on the other corner. Find the angles formed by Main and Water Streets.
1.7	The circular area around a flag pole has a diameter of 12 feet. A landscaper charges $2 a square foot to establish a lawn. How much would it cost to establish a lawn around the flag pole? Use 3.14 for π.

Home Involvement Activity

You Will Need: A tape measure
Directions: Measure the dimensions of a room in your home. Find the area. Find a type of carpet you like and find out how much the carpet costs per square foot. Not including labor for installation, how much would it cost to carpet the room?

Answers

1.1: Each number after the first two is the sum of the two numbers before it. The next number is 52. **1.2:** *Sample answer:* where the floor and one of the walls meet. **1.3:** Yes, both have length $\sqrt{29}$. **1.4:** acute; obtuse **1.5:** $(17, -7)$ **1.6:** $60°$ and $120°$ **1.7:** $226.08

CHAPTER 1

Prerequisite Skills Review
For use before Chapter 1

EXAMPLE 1 Subtracting Integers

Find the difference.

a. $-8 - 4$ b. $17 - (-11)$ c. $-\frac{3}{2} - \left(-\frac{1}{2}\right)$

SOLUTION

a. $-8 - 4 = -8 + (-4)$ Add the opposite of 4.
$= -12$ Use rules of addition.

b. $17 - (-11) = 17 + 11$ Add the opposite of -11.
$= 28$ Use rules of addition.

c. $-\frac{3}{2} - \left(-\frac{1}{2}\right) = -\frac{3}{2} + \frac{1}{2}$ Add the opposite of $-\frac{1}{2}$.
$= -1$ Use rules of addition.

Exercises for Example 1

Find the difference.

1. $9 - (-4)$ 2. $-6 - 8$ 3. $12 - (-5)$
4. $-7 - (-6)$ 5. $13 - 6$ 6. $5 - 14$

EXAMPLE 2 Evaluating Sums

Do all exponent work first, then add.

a. $3^2 + 5^2$ b. $(-2)^2 + 1^2$ c. $(-3)^3 + (-5)^2$

SOLUTION

a. $3^2 + 5^2 = 9 + 25$ Evaluate each power first.
$= 34$ Add.

b. $(-2)^2 + 1^2 = 4 + 1$ Evaluate each power first.
$= 5$ Add.

c. $(-3)^3 + (-5)^2 = -27 + 25$ Evaluate each power first.
$= -2$ Add.

CHAPTER 1 CONTINUED

NAME _____ DATE _____

Prerequisite Skills Review

For use before Chapter 1

Exercises for Example 2

Evaluate the sum.

7. $6^2 + 2^2$
8. $(-5)^2 + (-4)^2$
9. $(-2)^3 + 7^2$
10. $8^2 + (-3)^2$
11. $0^3 + (-6)^2$
12. $9^2 + (-1)^2$

EXAMPLE 3 Evaluating Radical Expressions

Evaluate each radical expression. Round your answer to two decimal places.

a. $\sqrt{49 + 16}$
b. $\sqrt{81 + 36}$

SOLUTION

a. $\sqrt{49 + 16} = \sqrt{65}$ Simplify.
   ≈ 8.06 Use a calculator.
b. $\sqrt{81 + 36} = \sqrt{117}$ Simplify.
   ≈ 10.82 Use a calculator.

Exercises for Example 3

Evaluate the radical expression. Round your answer to two decimal places.

13. $\sqrt{4 + 9}$
14. $\sqrt{36 + 25}$
15. $\sqrt{169 + 81}$
16. $\sqrt{64 + 49}$
17. $\sqrt{16 + 144}$
18. $\sqrt{100 + 121}$

CHAPTER 1

NAME _____ DATE _____

Strategies for Reading Mathematics

For use with Chapter 1

Strategy: Reading Your Textbook

At the beginning of every lesson in your textbook, you will find two goals that identify what you should learn. Every concept in the lesson is related to one of the two goals.

Goal 1 and **Goal 2** are clearly stated in the left margin on the first page of each lesson.

A practical application of the goals of the lesson is stated in the **Why you should learn it** statement. An example, exercise, or group of exercises that show the application are given.

> **What you should learn**
>
> **Goal 1** Understand and use the basic undefined terms and defined terms of geometry.
>
> **Goal 2** Sketch the intersections of lines and planes.
>
> **Why you should learn it**
>
> To name and draw the basic elements of geometry, including lines that intersect, as in the perspective drawing in **Exs. 68–72**.

The instructional part of each lesson is organized into two sections, one for each goal. The goal is briefly summarized in the head for the section and then is developed through explanations and examples in the section. Carefully read through all parts of the explanation under each goal in a lesson so that you will be ready to work the exercises.

> **STUDY TIP**
>
> **Using Student Help**
>
> Don't forget to read the Student Help notes found on the left-hand side of some pages. There are different uses for different types of Student Help notes. For example, the study tips help you avoid common errors and the skills review directs you to the pages in your textbook that review concepts you have studied in earlier math classes.

Questions

1. What should you learn in Lesson 1.2?

2. On the first page of a lesson, where can you look to find an application of the mathematical concepts you learn in that lesson? Name an application of the concepts in Lesson 1.2.

3. Name two different types of Student Help notes and tell what they help you do.

Strategies for Reading Mathematics

For use with Chapter 1

Visual Glossary

The Study Guide on page 2 lists the key vocabulary for Chapter 1. Use the visual glossary below to help you understand some of the key vocabulary in Chapter 1. You may want to copy these diagrams into your notebook and refer to them as you complete the chapter.

GLOSSARY

line segment (p. 11) Part of a line that consists of two points, called *endpoints*, and all points on the line that are between the endpoints. Also called *segment*.

ray (p. 11) Part of a line that consists of a point, called an *initial point*, and all points on the line that extend in one direction.

angle (p. 26) Consists of two different rays that have the same initial point.

bisect (p. 34) To divide into two congruent parts.

midpoint (p. 34) The point that divides, or bisects, a segment into two congruent segments.

complementary angles (p. 46) Two angles whose measures have the sum of 90°.

supplementary angles (p. 46) Two angles whose measures have the sum of 180°.

Angle and Segment Bisectors

A **segment bisector** is a segment, line, or ray that intersects a segment at its midpoint.

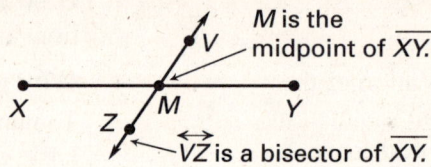

M is the midpoint of $\overline{XY}$.

$\overleftrightarrow{VZ}$ is a bisector of $\overline{XY}$.

An **angle bisector** is a ray that divides an angle into two adjacent angles that are congruent.

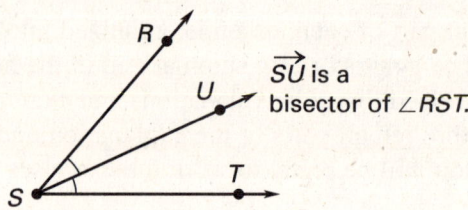

$\overrightarrow{SU}$ is a bisector of $\angle RST$.

Complementary and Supplementary Angles

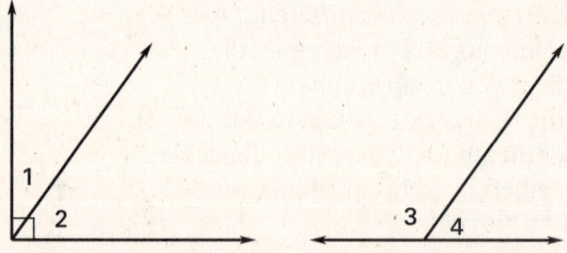

$m\angle 1 + m\angle 2 = 90°$ so $\angle 1$ and $\angle 2$ are complements of each other.

$m\angle 3 + m\angle 4 = 180°$ so $\angle 3$ and $\angle 4$ are supplements of each other.

LESSON 1.1

Teacher's Name _____ Class _____ Room _____ Date _____

Lesson Plan

1-day lesson (See *Pacing the Chapter,* TE pages 1C–1D)   For use with pages 3–9

GOALS
1. Find and describe patterns.
2. Use inductive reasoning to make real-life conjectures.

State/Local Objectives _____

_____

✓ Check the items you wish to use for this lesson.

STARTING OPTIONS
____ Prerequisite Skills Review: CRB pages 5–6
____ Strategies for Reading Mathematics: CRB pages 7–8
____ Warm-Up: TE page 3, CRB page 11, or Transparencies

TEACHING OPTIONS
____ Lesson Opener (Calculator): CRB page 12 or Transparencies
____ Technology Activity with Keystrokes: CRB page 13
____ Examples 1–6: SE pages 3–5
____ Extra Examples: TE pages 4–5 or Transparencies
____ Closure Question: TE page 5
____ Guided Practice Exercises: SE page 6

APPLY/HOMEWORK
Homework Assignment
____ Basic   12–26 even, 27, 28–46 even, 47, 48, 52–70 even
____ Average 12–26 even, 27, 28–46 even, 47, 48, 52–70 even
____ Advanced 12–26 even, 27, 28–46 even, 47–51, 52–70 even

Reteaching the Lesson
____ Practice Masters: CRB pages 14–16 (Level A, Level B, Level C)
____ Reteaching with Practice: CRB pages 17–18 or Practice Workbook with Examples
____ Personal Student Tutor

Extending the Lesson
____ Applications (Real-Life): CRB page 20
____ Challenge: SE page 9; CRB page 21 or Internet

ASSESSMENT OPTIONS
____ Checkpoint Exercises: TE pages 4–5 or Transparencies
____ Daily Homework Quiz (1.1): TE page 9, CRB page 24, or Transparencies
____ Standardized Test Practice: SE page 9; TE page 9; STP Workbook; Transparencies

Notes _____

_____

_____

LESSON 1.1

TEACHER'S NAME _____ CLASS _____ ROOM _____ DATE _____

Lesson Plan for Block Scheduling
Half-day lesson (See *Pacing the Chapter,* TE pages 1C–1D)

For use with pages 3–9

GOALS
1. Find and describe patterns.
2. Use inductive reasoning to make real-life conjectures.

State/Local Objectives _____
_____
_____

CHAPTER PACING GUIDE	
Day	Lesson
1	**1.1 (all)**; 1.2 (all)
2	1.3 (all)
3	1.4 (all); 1.5 (begin)
4	1.5 (end); 1.6 (begin)
5	1.6 (end); 1.7 (begin)
6	1.7 (end); Review Ch. 1
7	Assess Ch. 1; 2.1 (all)

✓ **Check the items you wish to use for this lesson.**

STARTING OPTIONS
____ Prerequisite Skills Review: CRB pages 5–6
____ Strategies for Reading Mathematics: CRB pages 7–8
____ Warm-Up: TE page 3, CRB page 11, or Transparencies

TEACHING OPTIONS
____ Lesson Opener (Calculator): CRB page 12 or Transparencies
____ Technology Activity with Keystrokes: CRB page 13
____ Examples 1–6: SE pages 3–5
____ Extra Examples: TE pages 4–5 or Transparencies
____ Closure Question: TE page 5
____ Guided Practice Exercises: SE page 6

APPLY/HOMEWORK
Homework Assignment (See also the assignment for Lesson 1.2.)
____ Block Schedule: 12–26 even, 27, 28–46 even, 47, 48, 52–70 even

Reteaching the Lesson
____ Practice Masters: CRB pages 14–16 (Level A, Level B, Level C)
____ Reteaching with Practice: CRB pages 17–18 or Practice Workbook with Examples
____ Personal Student Tutor

Extending the Lesson
____ Applications (Real-Life): CRB page 20
____ Challenge: SE page 9; CRB page 21 or Internet

ASSESSMENT OPTIONS
____ Checkpoint Exercises: TE pages 4–5 or Transparencies
____ Daily Homework Quiz (1.1): TE page 9, CRB page 24, or Transparencies
____ Standardized Test Practice: SE page 9; TE page 9; STP Workbook; Transparencies

Notes _____
_____
_____

Geometry
Chapter 1 Resource Book

LESSON 1.1

WARM-UP EXERCISES

For use before Lesson 1.1, pages 3–9

Evaluate each expression for the indicated value of x.

1. x^2, for $x = 0.5, 1, -1$

2. $(x + 1)(x - 1)$, for $x = 1, 2, 3$

3. $\dfrac{(x + 1)}{2}$, for $x = 1, 2, 3$

4. $\sqrt{x}$, for $x = \dfrac{1}{4}, 4, 9$

LESSON 1.1

NAME _____ DATE _____

Calculator Activity Lesson Opener

For use with pages 3–9

1. Use a calculator to evaluate the expressions.

    $9^2 = $ _____

    $99^2 = $ _____

    $999^2 = $ _____

2. Look for a pattern in Question 1 and use it to predict the value of the expressions. Then use a calculator to check your predictions.

    $9999^2 = $ _____

    $99,999^2 = $ _____

3. Use a calculator to evaluate the expressions.

    $12{,}345 \cdot 9 = $ _____

    $12{,}345 \cdot 18 = $ _____

    $12{,}345 \cdot 27 = $ _____

4. Look for a pattern in Question 3 and use it to predict the value of the expressions. Then use a calculator to check your predictions.

    $12{,}345 \cdot 36 = $ _____

    $12{,}345 \cdot 45 = $ _____

5. Use a calculator to evaluate the expressions.

    $3^2 = $ _____

    $33^2 = $ _____

    $333^2 = $ _____

Geometry
Chapter 1 Resource Book

Calculator Activity Lesson Opener

LESSON 1.1 CONTINUED

For use with pages 3–9

6. Look for a pattern in Question 5 and use it to predict the value of the expressions. Then use a calculator to check your predictions.

$3333^2 =$ _____

$33,333^2 =$ _____

Lesson 1.1 Practice A

For use with pages 3–9

Sketch the next figure in the pattern.

1.

2.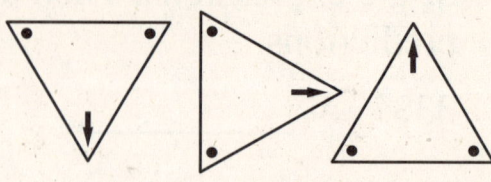

3.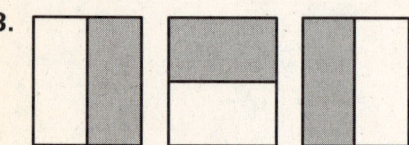

4.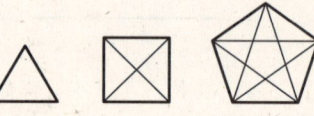

Describe a pattern in the sequence of numbers. Predict the next number.

5. 2, 5, 8, 11, . . .

6. 27, 9, 3, 1, . . .

7. 123, 234, 345, 456, . . .

8. 5, 7, 11, 17, 25, . . .

9. $\frac{1}{2}, \frac{2}{3}, \frac{3}{4}, \frac{4}{5}, \ldots$

10. $\frac{5}{4}, \frac{4}{6}, \frac{3}{8}, \frac{2}{10}, \ldots$

11. 4, 1, −2, −5, . . .

12. 1, 4, 9, 16, . . .

The first three objects in a pattern are shown. How many squares are in the next object?

13.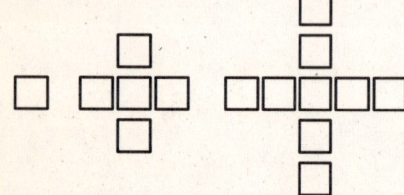

14.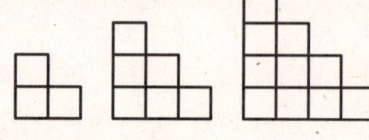

Complete the conjecture based on the pattern you observe in the specific cases.

15. The product of an odd number and an even number is ___?___ .

    3 · 8 = 24   6 · 5 = 30
    11 · 24 = 264   102 · 31 = 3162

16. The sum of an odd number and an even number is ___?___ .

    17 + 22 = 39   8 + 37 = 45
    135 + 48 = 183   94 + 85 = 179

Lesson 1.1 Practice B

For use with pages 3–9

Sketch the next figure in the pattern.

1.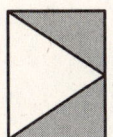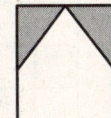

2.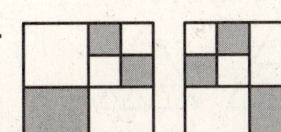

3.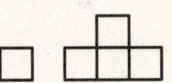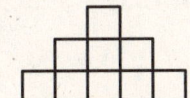

4.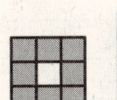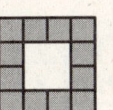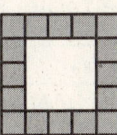

Describe a pattern in the sequence of numbers. Predict the next number.

5. 113, 224, 335, 446, . . .

6. 5, 7, 10, 14, 19, . . .

7. $\dfrac{1}{2}, \dfrac{3}{3}, \dfrac{5}{4}, \dfrac{7}{5}, \ldots$

8. $\dfrac{5}{6}, \dfrac{4}{5}, \dfrac{3}{4}, \dfrac{2}{3}, \ldots$

9. 4, 0, −4, −8, . . .

10. 4, 9, 16, 25, . . .

11. 2, 5, 11, 23, . . .

12. 2, 5, 11, 20, 32, . . .

The first three objects in a pattern are shown. How many squares are in the next object?

13.

14.

Show the conjecture is false by finding a counterexample.

15. The quotient of two whole numbers is a whole number.

16. The difference of the absolute value of two numbers is positive, meaning $|a| - |b| > 0$.

17. If $m \neq -1$, then $\dfrac{m}{m+1} < 1$.

Lesson 1.1 Practice C

For use with pages 3–9

Sketch the next figure in the pattern.

1.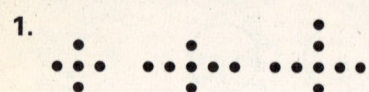

2.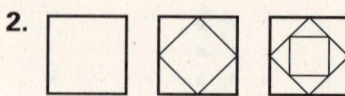

3.

4.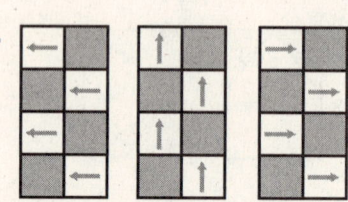

Describe a pattern in the sequence of numbers. Predict the next number.

5. 123, 133, 113, 123, . . .

6. 5, 8, 13, 20, 29, . . .

7. 22, 20, 17, 13, . . .

8. 0.49, 0.64, 0.81, 1, . . .

9. 8, 15, 29, 57, . . .

10. $-7, -4, -1, 2, \ldots$

11. $2, -4, 8, -16, \ldots$

12. $\frac{3}{7}, \frac{6}{5}, \frac{9}{3}, \frac{12}{1}, \ldots$

In Exercises 13 and 14, the number of bacteria after *n* hours is given in the table. Predict the number of bacteria after 8 hours.

13.
n (hours)	1	2	3	4	5
number of bacteria	3	6	12	24	48

14.
n (hours)	1	2	3	4	5
number of bacteria	640	320	160	80	40

In Exercises 15–17, show the conjecture is false by finding a counterexample.

15. The difference of two whole numbers is a whole number.

16. The absolute value of the sum of two numbers is the sum of their absolute values, meaning $|a + b| = |a| + |b|$.

17. If $m \neq -1$, then $\frac{m}{m - 1} > 1$.

LESSON 1.1

NAME _____ DATE _____

Reteaching with Practice

For use with pages 3–9

GOAL Find and describe patterns and use inductive reasoning

> **VOCABULARY**
>
> A **conjecture** is an unproven statement that is based on observations.
>
> **Inductive reasoning** is a process that involves looking for patterns and making conjectures.
>
> A **counterexample** is an example that shows a conjecture is false.

EXAMPLE 1 Describing a Visual Pattern

Sketch the next figure in the pattern.

 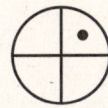 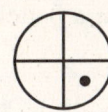

SOLUTION

Each figure looks like the one before it except that it has rotated 90°. The next figure will have the smaller circle in the lower-left quarter of the bigger circle.

Exercise for Example 1

1. Sketch the next figure in the pattern.

  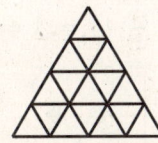

EXAMPLE 2 Describing a Number Pattern

Describe a pattern in the sequence of numbers. Predict the next number.

a. 5, 3, 1, −1, . . .   **b.** 1, −4, 9, −16, . . .   **c.** $\frac{1}{2}, \frac{1}{4}, \frac{1}{8}, \ldots$

SOLUTION

a. These are consecutive odd numbers, but listed backwards starting with 5. The next number is −3.

b. These numbers look like consecutive perfect squares, except that every other one is negative. The next number is 25.

c. Each number is $\frac{1}{2}$ times the previous number. The next number is $\frac{1}{16}$.

LESSON 1.1 CONTINUED

NAME _____ DATE _____

Reteaching with Practice

For use with pages 3–9

Exercises for Example 2

Describe a pattern in the sequence of numbers. Predict the next number.

2. 1, 2, 6, 24, . . .   **3.** 0, 3, 8, 15, 24, . . .

EXAMPLE 3 Making a Conjecture

Complete the conjecture.

Conjecture: The product of two consecutive even integers is divisible by __?__.

SOLUTION

List some specific examples and look for a pattern.

Examples:

$2 \times 4 = 8 = 8 \times 1$ $6 \times 8 = 48 = 8 \times 6$ $10 \times 12 = 120 = 8 \times 15$

$4 \times 6 = 24 = 8 \times 3$ $8 \times 10 = 80 = 8 \times 10$ $12 \times 14 = 168 = 8 \times 21$

Conjecture: The product of two consecutive even integers is divisible by 8.

Exercise for Example 3

Complete the conjecture based on the pattern you observe in the specific cases.

4. Conjecture: For any two numbers a and b, the product of $(a + b)$ and $(a - b)$ is always equal to __?__.

$(2 + 1) \times (2 - 1) = 3 = 2^2 - 1^2$ $(4 + 2) \times (4 - 2) = 12 = 4^2 - 2^2$

$(3 + 2) \times (3 - 2) = 5 = 3^2 - 2^2$ $(6 + 3) \times (6 - 3) = 27 = 6^2 - 3^2$

EXAMPLE 4 Finding a Counterexample

Show the conjecture is false by finding a counterexample.

Conjecture: All odd numbers are prime.

SOLUTION

The conjecture is false. Here is a counterexample: The number 9 is odd and is a composite number, not a prime number.

Exercise for Example 4

Show the conjecture is false by finding a counterexample.

5. The square of the sum of two numbers is equal to the sum of the squares of the two numbers. That is, $(a + b)^2 = a^2 + b^2$.

LESSON 1.1

NAME _____ DATE _____

Quick Catch-Up for Absent Students
For use with pages 3–9

The items checked below were covered in class on (date missed) _____

Lesson 1.1: Patterns and Inductive Reasoning

____ **Goal 1:** Find and describe patterns. (p. 3)

Material Covered:

____ Example 1: Describing a Visual Pattern

____ Example 2: Describing a Number Pattern

____ **Goal 2:** Use inductive reasoning to make real-life conjectures. (pp. 4–5)

Material Covered:

____ Example 3: Making a Conjecture

____ Example 4: Finding a Counterexample

____ Example 5: Examining an Unproven Conjecture

____ Example 6: Using Inductive Reasoning in Real Life

Vocabulary:

conjecture, p. 4   inductive reasoning, p. 4
counterexample, p. 4

____ Other (specify) _____
_____

Homework and Additional Learning Support

____ Textbook (specify) pp. 6–9 _____
_____

____ *Reteaching with Practice* worksheet (specify exercises) _____

____ *Personal Student Tutor* for Lesson 1.1

LESSON 1.1

NAME _____ DATE _____

Real-Life Application: When Will I Ever Use This?

For use with pages 3–9

Number Theory

Number Theory is a branch of mathematics that explores patterns in numbers. The mathematician Carl Friedrich Gauss (1777–1855) wrote, "Mathematics is the queen of the sciences and number theory the queen of mathematics." A story about Gauss's childhood illustrates his mathematical ability at an early age. In order to keep his students occupied, Gauss's teacher instructed them to add the numbers from one to one hundred and place their slates on his desk as soon as the task was completed. Right away, Gauss had the correct answer. How did he complete this seemingly difficult task so quickly?

1. Examine the numbers one to ten below.

1	2	3	4	5	6	7	8	9	10

   To find their sum, you could simply add from left to right, or you could pair the numbers as illustrated and add each pair.

1	2	3	4	5
10	9	8	7	6

   a. What is the sum of each pair?

   b. How many pairs are there?

2. Using the information in Exercise 1, what is the sum of the numbers one to ten?

3. Explain how you could quickly compute the sum of the numbers from one to twenty. What is the sum?

4. Generate a formula to compute the sum of the numbers from one to any number n. (*Hint:* Think of what each pair would add up to and how many pairs there would be.)

5. What was the answer to Gauss's problem?

6. Suppose you only wanted to add the even numbers from two to twenty. How would that change the formula you generated in Exercise 4?

LESSON 1.1

NAME _____ DATE _____

Challenge: Skills and Applications

For use with pages 3–9

In Exercises 1–4, consider the sequence 1, 9, 25, 49, ...

1. Describe a pattern in this sequence of numbers, and predict the next three numbers.

2. If any of these numbers is divided by 4, what is the remainder?

3. Based on your answer to part (c), write a conjecture.

4. Use algebra to explain why your conjecture is true.

In Exercises 5 and 6, consider the sequence of *triangular numbers*, which begins 1, 3, 6, 10, ...

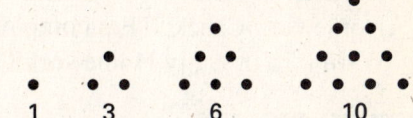

5. Describe the pattern, and predict the next five numbers.

6. Conjecture an algebraic formula for T_n, the nth triangular number. (*Hint:* Consider $1 \cdot 2, 2 \cdot 3, 3 \cdot 4$, and so on.)

In Exercises 7–11, suppose the interior of a circle is split into separate regions by drawing n line segments (called *chords*), each of which connects two points on the circle. The chords are chosen to maximize the number of regions formed.

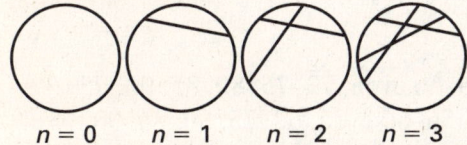

7. Find the number of regions for $n = 0, 1, 2,$ and 3.

8. How would you draw the fourth chord to maximize the number of regions? By how many regions will the number of regions increase when you draw the fourth chord?

9. By how many regions will the number of regions increase when you draw the nth chord?

10. Conjecture a sum of integers that gives the number of regions created by n chords.

11. Conjecture an algebraic formula for R_n, the number of regions created by n chords. (*Hint:* Compare the number of regions to the nth triangular number.)

LESSON 1.2

Teacher's Name _____ Class _____ Room _____ Date _____

Lesson Plan

1-day lesson (See *Pacing the Chapter*, TE pages 1C–1D)   For use with pages 10–16

GOALS
1. Understand and use the basic undefined terms of geometry.
2. Sketch the intersections of lines and planes.

State/Local Objectives _____

_____

✓ Check the items you wish to use for this lesson.

STARTING OPTIONS
____ Homework Check: TE page 6: Answer Transparencies
____ Warm-Up or Daily Homework Quiz: TE pages 10 and 9, CRB page 24, or Transparencies

TEACHING OPTIONS
____ Motivating the Lesson : TE page 11
____ Lesson Opener (Visual Approach): CRB page 25 or Transparencies
____ Examples 1–4: SE pages 10–12
____ Extra Examples: TE pages 11–12 or Transparencies; Internet
____ Closure Question: TE page 12
____ Guided Practice Exercises: SE page 13

APPLY/HOMEWORK
Homework Assignment
____ Basic   9, 12, 15, and every 3rd problem through 42, 44–51, 56–66 even, 73–75, 80, 85, 90, 95
____ Average 9, 12, 15, and every 3rd problem through 42, 44–51, 56–66 even, 68–75, 80, 85, 90, 95
____ Advanced 9, 12, 15, and every 3rd problem through 42, 44–51, 56–66 even, 68–76, 80, 85, 90, 95

Reteaching the Lesson
____ Practice Masters: CRB pages 26–28 (Level A, Level B, Level C)
____ Reteaching with Practice: CRB pages 29–30 or Practice Workbook with Examples
____ Personal Student Tutor

Extending the Lesson
____ Applications (Interdisciplinary): CRB page 32
____ Challenge: SE page 16; CRB page 33 or Internet

ASSESSMENT OPTIONS
____ Checkpoint Exercises: TE pages 11–12 or Transparencies
____ Daily Homework Quiz (1.2): TE page 16, CRB page 36, or Transparencies
____ Standardized Test Practice: SE page 16; TE page 16; STP Workbook; Transparencies

Notes _____

_____

_____

Geometry
Chapter 1 Resource Book

LESSON 1.2

TEACHER'S NAME _____ CLASS _____ ROOM _____ DATE _____

Lesson Plan for Block Scheduling

Half-day lesson (See *Pacing the Chapter*, TE pages 1C–1D) For use with pages 10–16

GOALS
1. Understand and use the basic undefined terms of geometry.
2. Sketch the intersections of lines and planes.

State/Local Objectives _____
_____
_____

CHAPTER PACING GUIDE	
Day	Lesson
1	1.1 (all); **1.2 (all)**
2	1.3 (all)
3	1.4 (all); 1.5 (begin)
4	1.5 (end); 1.6 (begin)
5	1.6 (end); 1.7 (begin)
6	1.7 (end); Review Ch. 1
7	Assess Ch. 1; 2.1 (all)

✓ **Check the items you wish to use for this lesson.**

STARTING OPTIONS
____ Homework Check: TE page 6: Answer Transparencies
____ Warm-Up or Daily Homework Quiz: TE pages 10 and 9, CRB page 24, or Transparencies

TEACHING OPTIONS
____ Motivating the Lesson : TE page 11
____ Lesson Opener (Visual Approach): CRB page 25 or Transparencies
____ Examples 1–4: SE pages 10–12
____ Extra Examples: TE pages 11–12 or Transparencies; Internet
____ Closure Question: TE page 12
____ Guided Practice Exercises: SE page 13

APPLY/HOMEWORK
Homework Assignment (See also the assignment for Lesson 1.1.)
____ Block Schedule: 9, 12, 15, and every 3rd problem through 42, 44–51, 56–66 even, 68–75, 80, 85, 90, 95

Reteaching the Lesson
____ Practice Masters: CRB pages 26–28 (Level A, Level B, Level C)
____ Reteaching with Practice: CRB pages 29–30 or Practice Workbook with Examples
____ Personal Student Tutor

Extending the Lesson
____ Applications (Interdisciplinary): CRB page 32
____ Challenge: SE page 16; CRB page 33 or Internet

ASSESSMENT OPTIONS
____ Checkpoint Exercises: TE pages 11–12 or Transparencies
____ Daily Homework Quiz (1.2): TE page 16, CRB page 36, or Transparencies
____ Standardized Test Practice: SE page 16; TE page 16; STP Workbook; Transparencies

Notes _____
_____
_____

LESSON 1.2

WARM-UP EXERCISES

For use before Lesson 1.2, pages 10–16

Give the coordinates of each point graphed below.

1. Point *A*
2. Point *B*
3. Point *C*
4. Point *D*

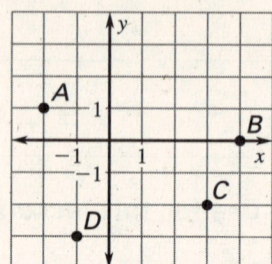

DAILY HOMEWORK QUIZ

For use after Lesson 1.1, pages 3–9

Describe a pattern in the sequence of numbers. Predict the next number.

1. 11, 23, 47, 95, . . .
2. 3, −11, −53, −179, . . .

Show the conjecture is false by finding a counterexample.

3. No two prime numbers are consecutive.
4. Division by an integer is always defined.

LESSON 1.2

NAME _____ DATE _____

Visual Approach Lesson Opener

For use with pages 10–16

You are using the map shown below to find your way around downtown.

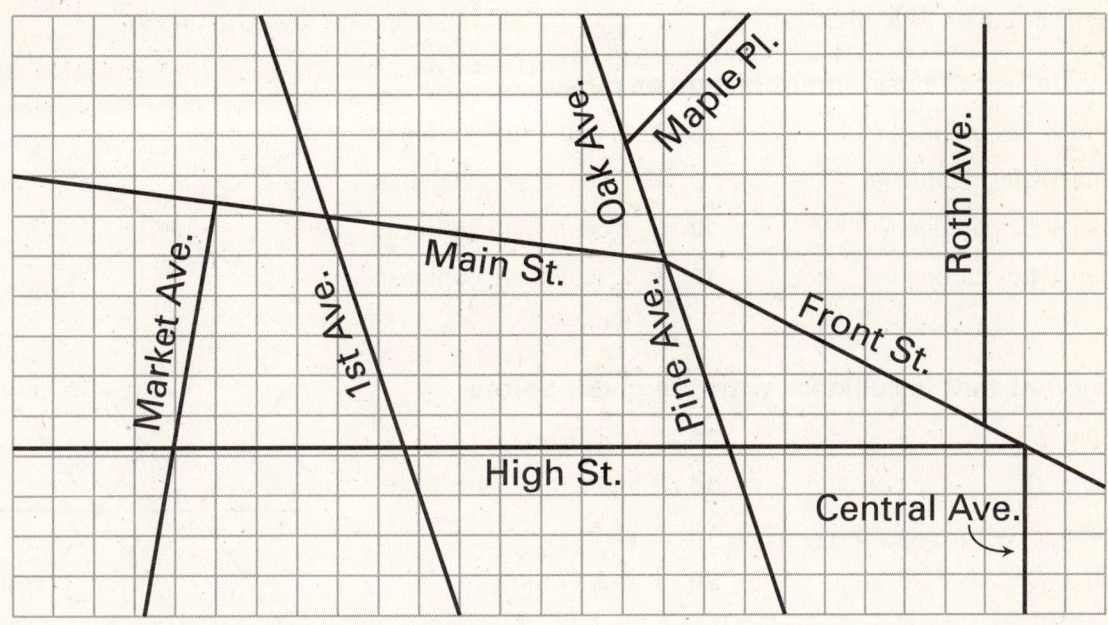

1. Locate Oak Avenue and Pine Avenue on the map. Make a dot to mark the intersection of Oak Avenue and Pine Avenue.

2. Make a dot to mark a different intersection on Oak Avenue than the one you marked in Question 1.

3. Make a dot to mark a different intersection on Pine Avenue than the one you marked in Question 1.

4. What appears to be true about the dots you made in Questions 1-3?

5. Locate Main Street and Front Street on the map. Make a dot to mark the intersection of Main Street and Front Street.

6. Make a dot to mark a different intersection on Main Street than the one you marked in Question 5.

7. Make a dot to mark a different intersection on Front Street than the one you marked in Question 5.

8. What appears to be true about the dots you made in Questions 5-7?

LESSON 1.2 Practice A
For use with pages 10–16

Draw a sketch and label as needed.

1. Three collinear points, *A*, *B*, and *C*.
2. $\overleftrightarrow{MN}$ intersecting $\overleftrightarrow{PQ}$ at point *R*.
3. Coplanar points *W*, *X*, *Y*, and *Z*.
4. Opposite rays, $\overrightarrow{JK}$ and $\overrightarrow{JC}$.

Decide whether the statement is *true* or *false*.

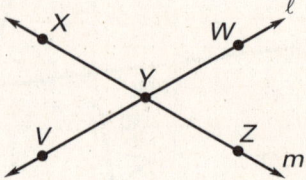

5. Point *X* lies on line *m*.
6. *X*, *Y*, and *Z* are collinear.
7. Point *W* lies on line *m*.
8. *X*, *Y*, and *Z* are coplanar.
9. Point *V* lies on line *l*.
10. *V*, *Y*, and *X* are collinear.
11. Point *Y* lies on line *l*.
12. *V*, *Y*, and *X* are coplanar.

Name a point that is collinear with the given points.

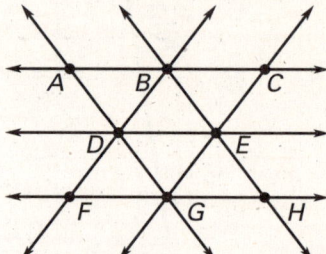

13. *B* and *E*
14. *F* and *H*
15. *D* and *G*
16. *A* and *C*
17. *H* and *E*
18. *G* and *C*
19. *A* and *D*
20. *B* and *C*

Name a point that is coplanar with the given points.

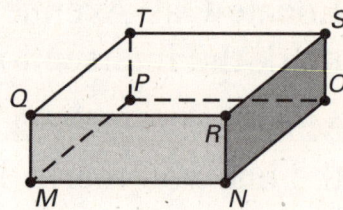

21. *M*, *N*, and *O*
22. *M*, *N*, and *R*
23. *T*, *Q*, and *M*
24. *T*, *Q*, and *R*
25. *T*, *S*, and *R*
26. *T*, *S*, and *O*
27. *O*, *S*, and *R*
28. *O*, *P*, and *M*

In Exercises 29–34, complete the sentence.

29. Collinear points are points that ___?___.
30. Coplanar points are points that ___?___.
31. $\overline{XY}$ consists of the endpoints *X* and *Y* and all points on the line $\overleftrightarrow{XY}$ that lie ___?___.
32. $\overrightarrow{MN}$ consists of the initial point *M* and all points on the line $\overleftrightarrow{MN}$ that lie ___?___.
33. Two rays or segments are collinear if they ___?___.
34. $\overrightarrow{PQ}$ and $\overrightarrow{PT}$ are opposite rays if ___?___.
35. Explain the difference between $\overrightarrow{BC}$ and $\overrightarrow{CB}$.

LESSON 1.2

Practice B
For use with pages 10–16

Decide whether the statement is *true* or *false*.

1. Point X lies on $\overrightarrow{ZY}$.
2. X, W, and Z are collinear.
3. Point W lies on $\overline{VY}$.
4. X, W, and Z are coplanar.
5. $\overrightarrow{YW}$ and $\overrightarrow{YV}$ are collinear.
6. $\overrightarrow{YW}$ and $\overrightarrow{YV}$ are coplanar.
7. $\overrightarrow{YX}$ and $\overrightarrow{YV}$ are collinear.
8. $\overrightarrow{YX}$ and $\overrightarrow{YV}$ are coplanar.

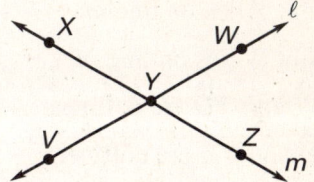

Name a point that is collinear with the given points.

9. B and E
10. C and H
11. D and G
12. A and C
13. H and E
14. G and B
15. B and I
16. B and C

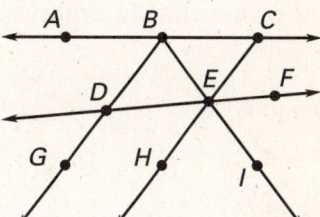

Name a point that is coplanar with the given points.

17. M, N, and R
18. M, N, and O
19. M, T, and Q
20. Q, T, and R
21. T, R, and S
22. Q, S, and O
23. O, P, and M
24. O, S, and R

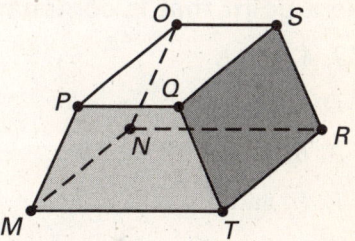

Complete the sentence.

25. $\overline{AB}$ consists of the endpoints A and B and all points on the line $\overleftrightarrow{AB}$ that lie ___?___.
26. $\overrightarrow{PQ}$ consists of the initial point P and all points on the line $\overleftrightarrow{PQ}$ that lie ___?___.
27. Two rays or segments are collinear if they ___?___.
28. $\overrightarrow{MN}$ and $\overrightarrow{ML}$ are opposite rays if ___?___.

Sketch the figure described.

29. Three points that are coplanar but not collinear.
30. Three lines that intersect at a single point.
31. A set of three lines that has two points of intersection.
32. A set of three lines that has three points of intersection.
33. Two planes that intersect.
34. Two planes that do not intersect.
35. Two rays that intersect at their initial points.
36. Two rays that do not intersect.

Lesson 1.2 Practice C

For use with pages 10–16

Decide whether the statement is *true* or *false*.

1. Point Y lies on line m.
2. X, Y, and Z are collinear.
3. Point W lies on line m.
4. X, Y, and Z are coplanar.
5. $\overrightarrow{YW}$ and $\overrightarrow{YD}$ are collinear.
6. $\overrightarrow{YW}$ and $\overrightarrow{YD}$ are coplanar.
7. $\overrightarrow{YX}$ and $\overrightarrow{YZ}$ are collinear.
8. $\overrightarrow{YX}$ and $\overrightarrow{YZ}$ are coplanar.

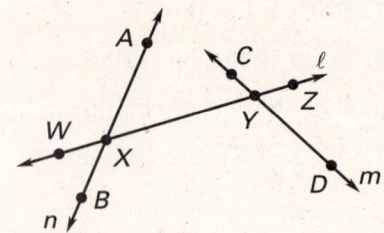

Name a point that is collinear with the given points.

9. B and E
10. F and E
11. D and G
12. A and C
13. G and E
14. F and C
15. A and D
16. B and C

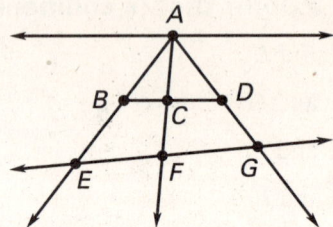

Name a point that is coplanar with the given points.

17. J, A, and E
18. B, C, and J
19. D, E, and A
20. H, E, and A
21. A, B, and H
22. A, B, and C
23. F, H, and B
24. D, E, and H

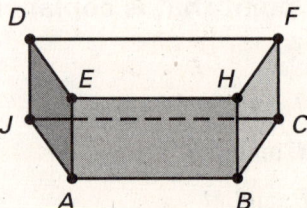

Sketch the lines, segments, and rays. Label your sketch.

25. Draw four noncollinear points A, B, C, and D. Then sketch $\overleftrightarrow{AB}$, $\overrightarrow{BC}$, $\overrightarrow{CD}$, and $\overline{DA}$.
26. Draw five noncollinear points M, N, O, P, and Q. Then sketch $\overline{MN}$, $\overrightarrow{OP}$, $\overrightarrow{PQ}$, $\overleftrightarrow{MP}$, and $\overline{NO}$.
27. Draw three collinear points H, I, and J with I between H and J. Add a point K between I and J.
28. Draw two points S and T. Then sketch $\overrightarrow{ST}$. Add a point U on the ray so that T is between S and U.

Sketch the figure described, if possible.

29. Three points that are collinear but not coplanar.
30. Three lines that intersect at a single point.
31. A set of four lines that has three points of intersection.
32. Three lines that do not intersect.
33. Two planes that intersect in a line.
34. Two planes that intersect in a single point.
35. Two rays that intersect at more than one point.
36. Two collinear rays that do not intersect.

LESSON 1.2

NAME _____ DATE _____

Reteaching with Practice

For use with pages 10–16

GOAL Understand and use the basic undefined terms and defined terms of geometry and sketch intersections of lines and planes

VOCABULARY

A **point** has no dimension, a **line** extends in one dimension, and a **plane** extends in two dimensions.

Collinear points are points that lie on the same line.

Coplanar points are points that lie on the same plane.

On a line passing through points A and B, **segment** AB consists of all points between A and B and **endpoints** A and B.

On a line passing through points A and B, **ray** AB consists of the **initial point** A and all points on the same side of A as point B.

If point C is between A and B, then ray CA and ray CB are **opposite rays.**

Two or more geometric figures **intersect** if they have one or more points in common. The **intersection** of the figures is the set of points the figures have in common.

EXAMPLE 1 Drawing and Naming Lines, Segments, and Rays

a. Draw three noncollinear points, A, B, and C. Then draw point D on line AB between points A and B. Draw segment CD. Draw ray CA and ray CB.

b. Are points A, B, and D collinear? Are points B, C, and D collinear?

c. Are ray CA and ray CB opposite rays? Are ray DA and ray DB opposite rays?

SOLUTION

a.

1. First, draw A, B, and C.   2. Draw line AB.   3. Draw D.

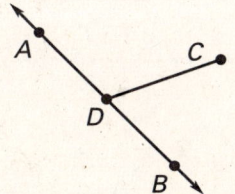

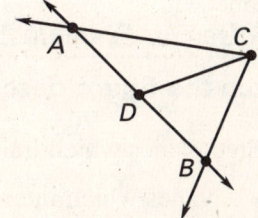

4. Draw segment CD.   5. Draw ray CA and ray CB.

Reteaching with Practice

For use with pages 10–16

b. Yes, points A, B, and D are collinear because they lie on line AB. No, points B, C, and D are noncollinear because a straight line cannot be drawn through all three points.

c. No, ray CA and ray CB are not opposite rays. Point C is not between A and B. Yes, ray DA and ray DB are opposite rays. Point D is between A and B.

Exercises for Example 1

1. Draw collinear points A, B, and C, with point B between A and C. Draw point D not on line AC. Draw line AD. Draw point E on line AD between point A and point D. Draw segment EC. Draw ray BD. Draw ray EB.

Use the diagram to name the figures.

2. Three noncollinear points
3. Two opposite rays
4. One line segment
5. Three collinear points
6. Two rays which are *not* opposite rays
7. Two line segments that are on the same line

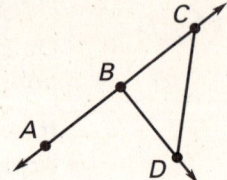

EXAMPLE 2 Sketching Intersections

Sketch the figure described.

a. Three lines that lie in the same plane, but two of the lines do not intersect with each other and the third line intersects with each of the other lines in a point.

b. Two planes which do not intersect, and a line which intersects each plane in a point.

SOLUTION

a.
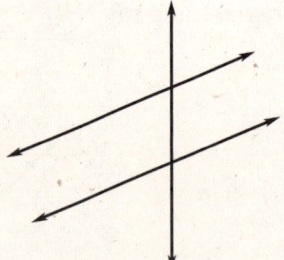

Draw two lines which do not intersect. Draw a third line, crossing each of the other lines.

b.
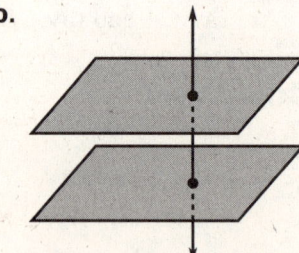

Draw two planes which do not intersect. Draw a line through both planes. Emphasize the points where the line intersects.

Exercises for Example 2

Sketch the figure described.

8. Three planes which intersect in a line

9. Two planes which intersect in a line, and a third plane which intersects each of the other two planes in a line, but not the same line

LESSON 1.2

NAME _____ DATE _____

Quick Catch-Up for Absent Students
For use with pages 10–16

The items checked below were covered in class on (date missed) _____

Lesson 1.2: Points, Lines, and Planes

____ **Goal 1:** Understand and use the basic undefined terms of geometry. (pp. 10–11)

Material Covered:

    ____ Example 1: Naming Collinear and Coplanar Points

    ____ Example 2: Drawing Lines, Segments, and Rays

    ____ Example 3: Drawing Opposite Rays

Vocabulary:

definition, p. 10	undefined terms, p. 10
point, p. 10	line, pp. 10, 11
plane, p. 10	collinear points, p. 10
coplanar points, p. 10	line segment, p. 11
endpoints, p. 11	ray, p. 11
initial point, p. 11	opposite rays, p. 11

____ **Goal 2:** Sketch the intersections of lines and planes. (p. 12)

Material Covered:

    ____ Activity: Modeling Intersections

    ____ Example 4: Sketching Intersections

Vocabulary:

    intersect, p. 12   intersection, p. 12

____ Other (specify) _____

_____

Homework and Additional Learning Support

    ____ Textbook (specify) pp. 13–16 _____

_____

    ____ Internet: Extra Examples at www.mcdougallittel.com

    ____ *Reteaching with Practice* worksheet (specify exercises) _____

    ____ *Personal Student Tutor* for Lesson 1.2

LESSON 1.2

NAME _____ DATE _____

Interdisciplinary Application

For use with pages 10–16

Latitude and Longitude

GEOGRAPHY In Geography class, you learn that Charleston, West Virginia and Savannah, Georgia are on the same imaginary vertical line, a line that extends from the North Pole to the South Pole called a longitude. Earth is divided into a grid using longitude lines and latitude lines (see the diagrams below). Latitudes are lines that circle the earth and are parallel to the equator. Latitude is measured in degrees north or south of the equator. The longitude line that runs through Greenwich, England is the 0° longitude line, or prime meridian. Longitude is measured in degrees east or west of the prime meridian.

Latitude

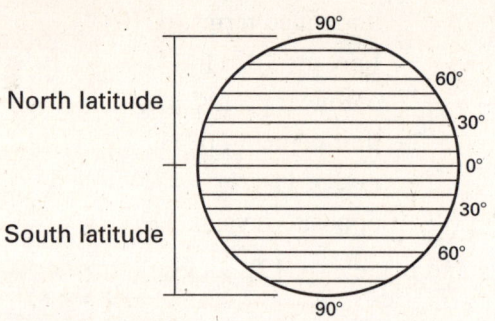

Longitude

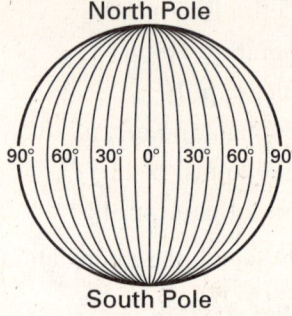

In Exercises 1–4, use the following table.

1. Using the table, locate other cities in the United States that are collinear with (on the same longitude as) Charleston, West Virginia and Savannah, Georgia.

2. Since longitude lines go from the top of Earth to the bottom and latitude lines go around, they intersect. More accurate locations for cities are given when they are pinpointed near an intersection of latitude and longitude. Find a city that is at the intersection point of 34° latitude and 118° longitude.

3. The capital of the United States is located at what intersection point of latitude and longitude?

4. If you left Los Angeles, California and hiked across country without veering off your latitude, in what state would you be when you arrived on the East Coast?

City	Latitude (in degrees)	Longitude (in degrees)
Albany, N.Y.	42	73
Anchorage, AK	61	149
Carlsbad, NM	32	104
Charleston, SC	32	79
Charleston, WV	38	81
Cleveland, OH	41	81
Columbia, SC	34	81
Fort Worth, TX	32	97
Jacksonville, FL	30	81
Key West, FL	24	81
Los Angeles, CA	34	118
New York City, NY	40	73
Philadelphia, PA	39	75
San Francisco, CA	37	122
Savannah, GA	32	81
St. Louis, MO	38	90
Virginia Beach, VA	36	75
Washington, D.C.	38	77

Lesson 1.2

Challenge: Skills and Applications
For use with pages 10–16

An example of a *two-point perspective drawing* is shown on page 15 of your textbook.

In a *one-point perspective drawing*, there is only one vanishing point. Horizontal lines going across the viewer's field of vision are shown as horizontal, and, as in a two-point perspective drawing, vertical lines are shown vertical. An example of a one-point perspective drawing is shown at the right.

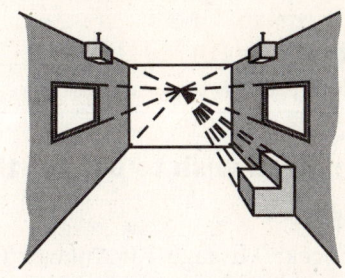

In Exercises 1–4, identify the drawing as *one-point perspective*, *two-point perspective*, or *neither*.

1.

2.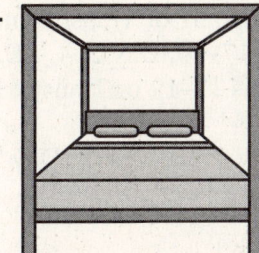

3.

4.

5. Create an original one-point perspective drawing and an original two-point perspective drawing. Identify which drawing is which.

LESSON 1.3

Teacher's Name _____ Class _____ Room _____ Date _____

Lesson Plan

2-day lesson (See *Pacing the Chapter*, TE pages 1C–1D)   For use with pages 17–25

 **GOALS** 1. Use segment postulates.
2. Use the Distance Formula to measure distances.

State/Local Objectives _____
_____

✓ **Check the items you wish to use for this lesson.**

STARTING OPTIONS
____ Homework Check: TE page 13: Answer Transparencies
____ Warm-Up or Daily Homework Quiz: TE pages 17 and 16, CRB page 36, or Transparencies

TEACHING OPTIONS
____ Motivating the Lesson : TE page 18
____ Lesson Opener (Activity): CRB page 37 or Transparencies
____ Examples: Day 1: 1–3: SE pages 17–20; Day 2: 4, SE page 20
____ Extra Examples: Day 1: TE pages 18–19 or Transp.; Day 2: TE page 20 or Transp.
____ Closure Question: TE page 20
____ Guided Practice: SE page 21   Day 1: Exs. 1–12; Day 2: Exs. none

APPLY/HOMEWORK
Homework Assignment
____ Basic   Day 1: 14–42 even; Day 2: 44–49, 55, 56, 60–71; Quiz 1: 1–7
____ Average   Day 1: 14–42 even; Day 2: 44–49, 53–56, 60–71; Quiz 1: 1–7
____ Advanced   Day 1: 14–42 even; Day 2: 44–49, 53–71; Quiz 1: 1–7

Reteaching the Lesson
____ Practice Masters: CRB pages 38–40 (Level A, Level B, Level C)
____ Reteaching with Practice: CRB pages 41–42 or Practice Workbook with Examples
____ Personal Student Tutor

Extending the Lesson
____ Cooperative Learning Activity: CRB page 44
____ Applications (Real-Life): CRB page 45
____ Math & History: SE page 25; CRB page 46; Internet
____ Challenge: SE page 24; CRB page 47 or Internet

ASSESSMENT OPTIONS
____ Checkpoint Exercises: Day 1: TE pages 18–19 or Transp.; Day 2: TE page 20 or Transp.
____ Daily Homework Quiz (1.3): TE page 24, CRB page 51, or Transparencies
____ Standardized Test Practice: SE page 24; TE page 24; STP Workbook; Transparencies
____ Quiz (1.1–1.3): SE page 25; CRB page 48

Notes _____
_____
_____

LESSON 1.3

Teacher's Name _____ **Class** _____ **Room** _____ **Date** _____

Lesson Plan for Block Scheduling
1-day lesson (See *Pacing the Chapter,* TE pages 1C–1D)   For use with pages 17–25

GOALS
1. Use segment postulates.
2. Use the Distance Formula to measure distances.

State/Local Objectives _____

_____

_____

CHAPTER PACING GUIDE	
Day	Lesson
1	1.1 (all); 1.2 (all)
2	**1.3 (all)**
3	1.4 (all); 1.5 (begin)
4	1.5 (end); 1.6 (begin)
5	1.6 (end); 1.7 (begin)
6	1.7 (end); Review Ch. 1
7	Assess Ch. 1; 2.1 (all)

✓ **Check the items you wish to use for this lesson.**

STARTING OPTIONS
___ Homework Check: TE page 13: Answer Transparencies
___ Warm-Up or Daily Homework Quiz: TE pages 17 and 16, CRB page 36, or Transparencies

TEACHING OPTIONS
___ Motivating the Lesson : TE page 18
___ Lesson Opener (Activity): CRB page 37 or Transparencies
___ Examples: 1–4, SE pages 17–20
___ Extra Examples: TE pages 18–20 or Transparencies
___ Closure Question: TE page 20
___ Guided Practice Exercises: SE page 21

APPLY/HOMEWORK
Homework Assignment
___ Block Schedule: 14–42 even, 44–49, 53–71; Quiz 1: 1–7

Reteaching the Lesson
___ Practice Masters: CRB pages 38–40 (Level A, Level B, Level C)
___ Reteaching with Practice: CRB pages 41–42 or Practice Workbook with Examples
___ Personal Student Tutor

Extending the Lesson
___ Cooperative Learning Activity: CRB page 44
___ Applications (Real-Life): CRB page 45
___ Math & History: SE page 25; CRB page 46; Internet
___ Challenge: SE page 24; CRB page 47 or Internet

ASSESSMENT OPTIONS
___ Checkpoint Exercises: TE pages 18–20 or Transparencies
___ Daily Homework Quiz (1.3): TE page 24, CRB page 51, or Transparencies
___ Standardized Test Practice: SE page 24; TE page 24; STP Workbook; Transparencies
___ Quiz (1.1–1.3): SE page 25; CRB page 48

Notes _____

_____

_____

LESSON 1.3

WARM-UP EXERCISES

For use before Lesson 1.3, pages 17–25

Evaluate each expression.

1. $|4 - 7|$

2. $|4.3 - 1.2|$

3. $\sqrt{4^2 + 3^2}$

4. $\sqrt{(-2)^2 + 3^2}$

DAILY HOMEWORK QUIZ

For use after Lesson 1.2, pages 10–16

Decide whether the statement is *true* or *false*.

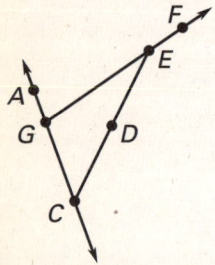

1. *A*, *G*, and *C* are collinear.

2. *A*, *G*, and *D* are coplanar.

3. *D* and *E* lie on a ray.

4. *G* lies on the intersection of two lines.

5. *C* lies on the intersection of a segment and a line.

LESSON 1.3

Activity Lesson Opener
For use with pages 17–25

SET UP: Work with a partner.

1. Points A, B, C, D, and E lie on the line below but not necessarily in that order. Use the clues to help you place each point in its correct position on the line.

   a. Point A is to the right of point B.
   b. Point C is between points A and E.
   c. Points B and E are 11 cm apart.
   d. Point D is between points B and A.
   e. Point E is at the 12 cm mark.
   f. Point C is 2 cm away from point E.
   g. Point D is 1 cm away from point B.
   h. Point A is 2 cm closer to point D than to point C.

Use your completed drawing to find the length of the segment.

2. $\overline{AD}$ 3. $\overline{AB}$ 4. $\overline{AC}$

5. $\overline{DE}$ 6. $\overline{CD}$ 7. $\overline{BC}$

LESSON 1.3

Practice A
For use with pages 17–25

Use a ruler to measure the length of each line segment to the nearest millimeter.

1.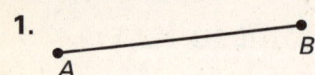
2.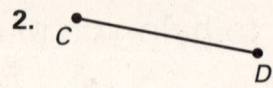
3.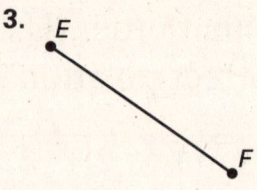
4.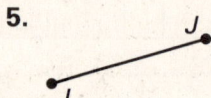
5.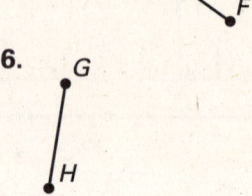
6. (see figure)

Draw a sketch of the three collinear points. Then write the Segment Addition Postulate for the points.

7. S is between D and P.
8. J is between S and H.
9. C is between Q and R.
10. T is between M and N.

In the diagram of collinear points, $GK = 24$, $HJ = 10$, and $GH = HI = IJ$. Find each length.

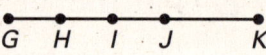

11. HI
12. IJ
13. GH
14. JK
15. IG
16. IK

Suppose J is between H and K. Use the Segment Addition Postulate to solve for x. Then find the length of each segment.

17. $HJ = 5x$
    $JK = 7x$
    $KH = 96$

18. $HJ = 2x + 5$
    $JK = 3x - 7$
    $KH = 18$

19. $HJ = 6x - 5$
    $JK = 4x - 6$
    $KH = 129$

Find the distance between each pair of points.

20. $A(3, 2)$, $B(2, 0)$
21. $C(1, 3)$, $D(-2, 4)$
22. $E(-1, 0)$, $F(2, -4)$

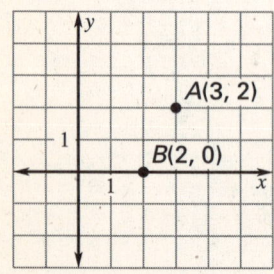
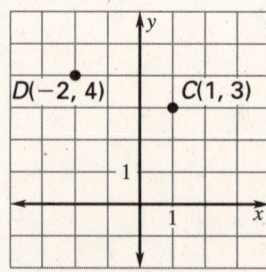
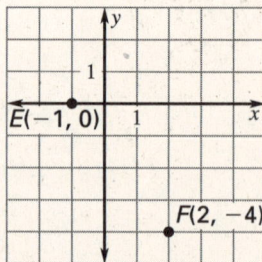

Use the Distance Formula to decide whether $\overline{AB} \cong \overline{BC}$.

23. $A(0, 1)$
    $B(2, 4)$
    $C(4, 7)$

24. $A(-3, 1)$
    $B(1, -1)$
    $C(6, -3)$

25. $A(4, 2)$
    $B(-1, -1)$
    $C(-6, -4)$

LESSON 1.3

Practice B
For use with pages 17–25

Use a ruler to measure the length of each line segment to the nearest millimeter.

1.

2. [segment from C to D]

3. [segment from E to F]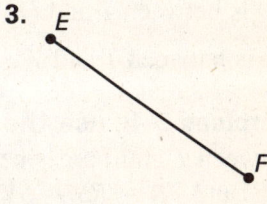

Draw a sketch of the three collinear points. Then write the Segment Addition Postulate for the points.

4. A is between T and Q.

5. M is between H and A.

6. J is between S and H.

7. A is between L and B.

In Exercises 8–11, use the following information.
S is between T and V. R is between S and T. T is between R and Q. $QV = 18$, $QT = 6$, and $TR = RS = SV$. Make a sketch and answer the following.

8. Find RS.
9. Find QS.
10. Find TS.
11. Find TV.

Suppose J is between H and K. Use the Segment Addition Postulate to solve for x. Then find the length of each segment.

12. $HJ = 2x + 4$
    $JK = 3x + 3$
    $KH = 22$

13. $HJ = 5x - 3$
    $JK = 8x - 9$
    $KH = 131$

14. $HJ = 2x + \frac{1}{3}$
    $JK = 5x + \frac{2}{3}$
    $KH = 12x - 4$

Find the distance between each pair of points.

15. $D(1, 3), E(-2, 4), F(0, -4)$

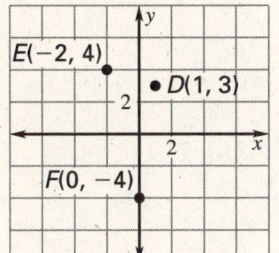

16. $G(-1, 0), H(2, -4), I(1, 3)$

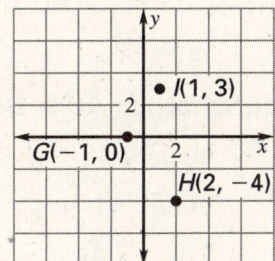

17. $A(3, 2), B(2, 0), C(1, -3)$

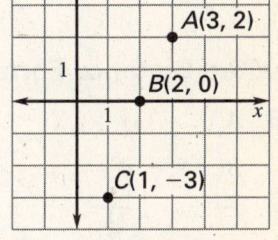

18. **Marathon** The map at the right is being used to plan a 26.3 mile marathon. Coordinates are given in miles. The locations of the participating towns on the map are: Curtis (0, 0), Clearfield (10, 2), Buster (5, 7), and Angel City (1, 4).

    Which of the following planned routes is nearest to the 26.3 mile requirement?

    (a) Curtis to Clearfield to Angel City to Curtis
    (b) Curtis to Clearfield to Buster to Angel City to Curtis
    (c) Curtis to Buster to Clearfield to Curtis
    (d) Curtis to Buster to Angel City to Clearfield to Curtis

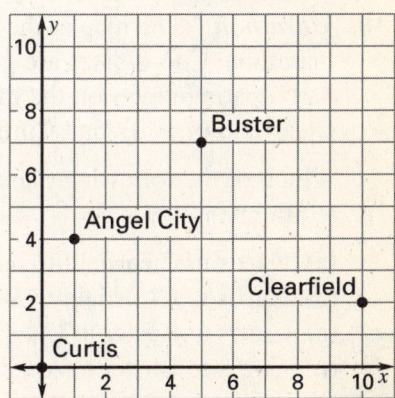

LESSON 1.3

NAME _____ DATE _____

Practice C
For use with pages 17–25

Draw a sketch of the three collinear points. Then write the Segment Addition Postulate for the points.

1. D is between T and Q.
2. M is between Q and N.
3. L is between T and W.
4. A is between X and Y.

In Exercises 5–8, use the following information.
S is between T and V. R is between S and T. T is between R and Q. $QV = 23$, $QT = 8$, and $TR = RS = SV$. Make a sketch and answer the following.

5. Find RS.
6. Find QS.
7. Find TS.
8. Find TV.

Suppose J is between H and K. Use the Segment Addition Postulate to solve for x. Then find the length of each segment.

9. $HJ = 3(x + 2)$
   $JK = 3x - 4$
   $KH = 44$

10. $HJ = 8x - 3$
    $JK = 12x - 5$
    $KH = 112$

11. $HJ = \frac{1}{3}x + 4$
    $JK = 2x + \frac{2}{3}$
    $KH = 2\frac{2}{3}x + 1$

Find the distance between each pair of points.

12. $D(-1, 3)$, $E(-2, 5)$, $F(0, -7)$
13. $G(-5, 0)$, $H(6, -2)$, $I(3, 6)$
14. $A(5, 3)$, $B(7, 0)$, $C(2, -4)$

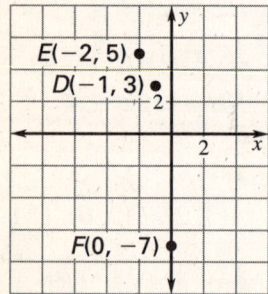
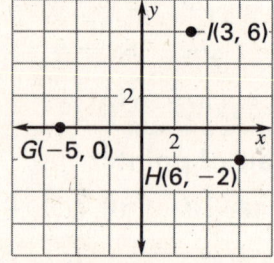
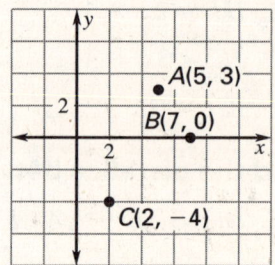

Use the Distance Formula to decide whether $\overline{AB} \cong \overline{BC}$.

15. $A(0, -1)$
    $B(-2, -4)$
    $C(-4, -7)$

16. $A(-3, 1)$
    $B(1.5, -1.5)$
    $C(6, -3.5)$

17. $A(4, 2)$
    $B(-1, -1)$
    $C(-6, -4)$

18. **Marathon** The map at the right is being used to plan a 26.3 mile marathon. Coordinates are given in miles. The locations of the participating towns on the map are: Angel City $(0, 0)$, Buster $(4, 3)$, Clearfield $(9, -2)$, and Curtis $(-1, -4)$.

    Which of the following planned routes is nearest to the 26.3 mile requirement?

    (a) Curtis to Clearfield to Angel City to Curtis
    (b) Curtis to Clearfield to Buster to Angel City to Curtis
    (c) Curtis to Buster to Clearfield to Curtis
    (d) Curtis to Buster to Angel City to Clearfield to Curtis

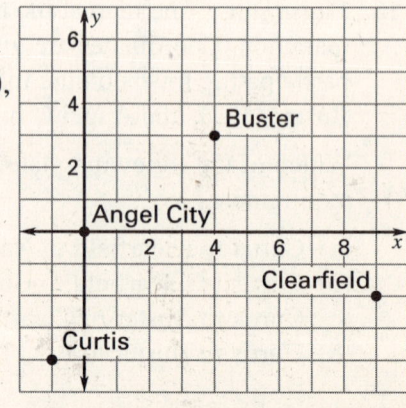

LESSON 1.3

NAME _____ DATE _____

Reteaching with Practice

For use with pages 17–25

GOAL Use segment postulates and use the distance formula to measure distances

VOCABULARY

A **postulate** or **axiom** is a rule that is accepted without proof.

Postulate 1 Ruler Postulate:
The points on a line can be matched one to one with the real numbers. The real number that corresponds to a point is the **coordinate** of the point.

The **distance** between points A and B, written as AB, is the absolute value of the difference between the coordinates of A and B.

AB is also called the **length** of $\overline{AB}$.

When three points lie on a line, you can say that one of them is **between** the other two.

Postulate 2 Segment Addition Postulate:
If B is between A and C, then $AB + BC = AC$. If $AB + BC = AC$, then B is between A and C.

The **Distance Formula** is a formula for computing the difference between two points in a coordinate plane.

The Distance Formula:
If $A(x_1, y_1)$ and $B(x_2, y_2)$ are points in a coordinate plane, then the distance between A and B is $AB = \sqrt{(x_2 - x_1)^2 + (y_2 - y_1)^2}$.

Segments that have the same length are called **congruent segments.**

EXAMPLE 1 Using the Segment Addition Postulate

In the diagram of the collinear points, $DE = 2$, $EF = 3$, and $DE = FG$. Find each length.
FG
DF
DG
EG

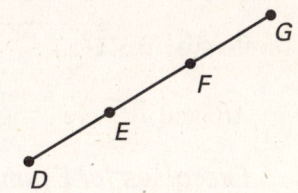

SOLUTION

Since $DE = FG$ and $DE = 2$, $FG = 2$.

Since $DF = DE + EF$, $DF = 2 + 3 = 5$.

Since $DG = DF + FG$, $DG = 5 + 2 = 7$.

Since $EG = EF + FG$, $EG = 3 + 2 = 5$.

LESSON 1.3 CONTINUED

NAME _____ DATE _____

Reteaching with Practice

For use with pages 17–25

Exercises for Example 1

1. In the diagram of the collinear points, $BC = 5$ and $BC = AB$. Find the following lengths.
   a. AC
   b. AB
   c. Are any segments congruent?

2. In the diagram of the collinear points, $HK = 9$, $HI = JK$, and $IJ = 1$. Find the following lengths.
   a. HI
   b. JK
   c. HJ
   d. IK

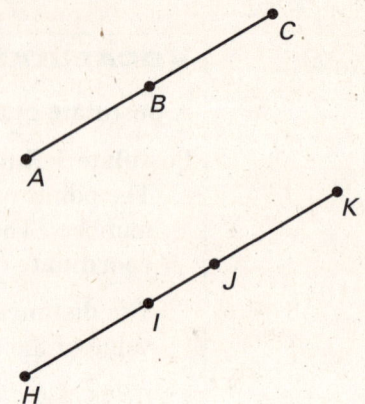

EXAMPLE 2 Using the Distance Formula

Find the following distances. State whether any of the segments are congruent.
   a. AB
   b. BC
   c. CD
   d. AC

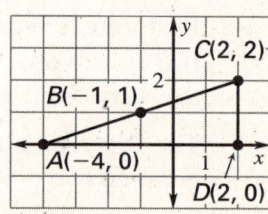

SOLUTION

Use the Distance Formula.

a. $AB = \sqrt{[(-1)-(-4)]^2 + (1-0)^2} = \sqrt{3^2 + 1^2} = \sqrt{9+1} = \sqrt{10}$

b. $BC = \sqrt{[2-(-1)]^2 + (2-1)^2} = \sqrt{3^2 + 1^2} = \sqrt{9+1} = \sqrt{10}$

c. $CD = \sqrt{(2-2)^2 + (0-2)^2} = \sqrt{0^2 + (-2)^2} = \sqrt{0+4} = \sqrt{4} = 2$

d. $AC = \sqrt{[2-(-4)]^2 + (2-0)^2} = \sqrt{6^2 + 2^2} = \sqrt{36+4} = \sqrt{40} = 2\sqrt{10}$

$\overline{AB}$ and $\overline{BC}$ are congruent because they have the same length.

Exercises for Example 2

Find the distance between the points whose coordinates are given.

3. $(6, 4), (-8, 11)$
4. $(-5, 8), (-10, 14)$
5. $(-4, -20), (-10, 15)$
6. $(40, 32), (36, 20)$
7. $(5, -8), (0, 0)$
8. $(a, b), (-a, -b)$

LESSON 1.3

NAME _____ DATE _____

Quick Catch-Up for Absent Students
For use with pages 17–25

The items checked below were covered in class on (date missed) _____

Lesson 1.3: Segments and Their Measures

____ **Goal 1:** Use segment postulates. (pp. 17–18)

Material Covered:

____ Example 1: Finding the Distance Between Two Points

____ Example 2: Finding Distances on a Map

Vocabulary:

    postulates, p. 17   axioms, p. 17
    coordinate, p. 17   distance, p. 17
    length, p. 17   between, p. 18

____ **Goal 2:** Use the Distance Formula to measure distances. (pp. 19–20)

Material Covered:

____ Student Help: Study Tip

____ Example 3: Using the Distance Formula

____ Student Help: Study Tip

____ Student Help: Study Tip

____ Example 4: Finding Distances on a City Map

Vocabulary:

    Distance Formula, p. 19   congruent segments, p. 19

____ Other (specify) _____
_____

Homework and Additional Learning Support

____ Textbook (specify) <u>pp. 21–25</u> _____
_____

____ *Reteaching with Practice* worksheet (specify exercises) _____

____ *Personal Student Tutor* for Lesson 1.3

LESSON 1.3

NAME _____ **DATE** _____

Cooperative Learning Activity

For use with pages 17–25

GOAL To apply the Distance Formula to decode a message

Materials: graph paper, loose-leaf paper, pencil

Exploring the Distance Formula

It is easy to find the distance between two points in a coordinate plane that form a vertical or horizontal segment using subtraction. The Distance Formula, however, provides a way to easily find the distance between any two points in a coordinate plane. In this activity, your group will apply the Distance Formula to encode and decode messages hidden in a coordinate plane.

Instructions

1. Use graph paper to make a coordinate plane. Graph and label enough points to spell out a word or short phrase. (Repeated letters should each have their own point.)

2. On a separate sheet of paper, write the coordinate of each point and use the Distance Formula to calculate the distance between each of the letters in the message. Make sure that no two points are the same distance apart.

3. Encode the message by writing the distances in order and identifying a starting point.

4. Exchange the encoded message with another group. Decode the other group's message, showing the calculation for each distance.

Analyzing the Results

1. Did your group have to use the Distance Formula to find all of the correct points? Explain.

2. Describe the strategy your group used to find the consecutive points in the message.

Sample: Start at S.

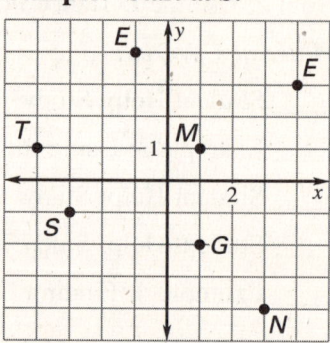

Code:
$\sqrt{29}, 2\sqrt{10}, 3, \sqrt{13}, 5\sqrt{2}, \sqrt{74}$

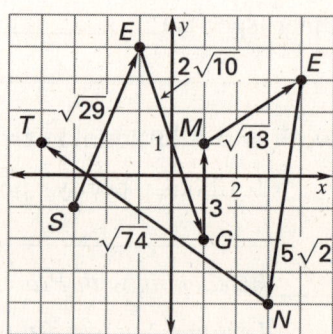

Answer: Segment

Real-Life Application: When Will I Ever Use This?

For use with pages 17–25

Astronomical Units

Astronomers use different units of measure to calculate the distances that the planets are from each other and from the sun. They can use miles, kilometers, and light years. Another unit they use is the astronomical unit, abbreviated AU. One astronomical unit is equal to the average distance between the sun and Earth, which is 92,960,000 miles.

1. In the table, the distances of the planets from the sun are listed in astronomical units. On a number line, place each planet according to its distance from the sun. Let the sun's position on the number line be at point zero.

Planet	Distance from the sun (in AUs)
Earth	1
Jupiter	5.20
Mars	1.52
Mercury	0.39
Neptune	30.06
Pluto	39.44
Saturn	9.54
Uranus	19.18
Venus	0.72

2. Which planets are between Earth and the sun?

3. Which planet is closest to Earth?

4. Recently, NASA sent a probe to Mars. How far is Mars from Earth?

5. Which two neighboring planets are furthest apart?

6. The distance between Earth and the moon is 240,000 miles. Calculate the moon's distance from Earth in astronomical units.

7. Because the moon rotates around Earth, it is sometimes between Earth and the sun (solar eclipse) and other times Earth is between the moon and the sun (lunar eclipse). Use Exercise 6 to position the moon on your number line with respect to Earth and the sun during a solar eclipse and during a lunar eclipse.

Lesson 1.3

Math and History Application
For use with page 25

HISTORY Ancient Greece produced many talented mathematicians, but Euclid (c. 300 B.C.) is in a class by himself. His *The Elements*, comprised of thirteen volumes, remains the ultimate geometry book more than two thousand years after its initial publication. Many results in *The Elements* were originally discovered by earlier mathematicians, but Euclid is responsible for their presentation. Compiling a wide variety of individual discoveries, Euclid was the first to arrange the results in a logical sequence, following deductively from a minimal set of initial postulates.

Euclid's achievements are even more remarkable when one considers the limited tools he had at his disposal: a **straightedge**, and a **compass.** It should be pointed out that the straightedge and compass Euclid used were different from the rulers and compasses of today. The straightedge had no markings, and the compass Euclid used would collapse if either leg were lifted off the paper. It was therefore not possible to measure a certain distance and then "transfer" that distance elsewhere.

MATH In Proposition 1 of *The Elements*, Volume I, Euclid used a straightedge and a compass to construct an **equilateral triangle**, a triangle that has three congruent sides.

1. Use a straightedge to draw a line segment *AB*.

2. Use a compass to draw a circle having center *A* and radius $\overline{AB}$.

3. Use a compass to draw a circle having center *B* and radius $\overline{AB}$.

4. Observe that the circles in Exercises 2 and 3 intersect at two points. Label these points *C* and *D*. Use a straightedge to draw the line segments *AC* and *BC*.

5. Use a ruler to verify that *ABC* is an equilateral triangle; that is, verify that
$$AB = BC = AC.$$

6. Discuss how you would use a straightedge and a compass to construct a **right triangle**, a triangle that has a 90° angle.

Lesson 1.3 Challenge: Skills and Applications

For use with pages 17–25

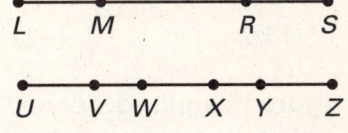

1. Suppose $LM = RS$. Use the Segment Addition Postulate to show that $LR = MS$.

2. Suppose $\overline{UV} \cong \overline{WX} \cong \overline{YZ}$ and $\overline{VW} \cong \overline{XY}$. What other segments must be congruent? (Use the Segment Addition Postulate.)

In Exercises 3 and 4, consider the following statement.

If A, B, and C are collinear points, then $AB + BC = AC$.

3. Explain how the statement differs from the first half of the Segment Addition Postulate.

4. Sketch a counterexample to show that the statement is false.

In Exercises 5–7, let A, B, C, and D be four points in a plane. Tell whether the given condition is sufficient to conclude that $AB + BC + CD = AD$. Justify your answer by using the Segment Addition Postulate or by sketching a counterexample.

5. B is between A and C, and B is between A and D.

6. B is between A and D, and C is between B and D.

7. B and C are both between A and D.

In Exercises 8–10, assume that K is between J and L.

8. If $JK = 2x + 5$, $KL = 5x + 3$, and $JL = x^2$, find x.
   (*Hint:* Make sure your answer is reasonable!)

9. If $JK = 20 - x^2$, $KL = 2 - x$, and $JL = 10$, find x.

10. If the ratio of KL to JK is 2:7, and $JL = 162$, find JK.

In Exercises 11–13, use the following information to determine whether $\overline{PQ} \cong \overline{QR}$.

In a three-dimensional coordinate system, the distance between two points (x_1, y_1, z_1) and (x_2, y_2, z_2) is

$$\sqrt{(x_2 - x_1)^2 + (y_2 - y_1)^2 + (z_2 - z_1)^2}.$$

11. $P(3, 7, -2)$
    $Q(2, 5, 6)$
    $R(5, 0, 4)$

12. $P(-4, 1, 2)$
    $Q(-1, -4, 4)$
    $R(1, -7, 9)$

13. $P(1, 0, -3)$
    $Q(2, -4, -1)$
    $R(0, -4, 4)$

LESSON 1.3 Quiz 1

NAME _____ DATE _____

For use after Lessons 1.1–1.3

Write the next number in the sequence. *(Lesson 1.1)*

1. 20, 18.5, 17, 15.5, ...
2. 0, 3, −3, 6, −6, ...

Sketch the figure described. *(Lesson 1.2)*

3. Two rays that have the same initial point A.

4. Two intersecting lines, and a third line that intersects both lines.

5. One line that intersects a line in a different plane.

6. Two planes that intersect.

7. **Around the Mountains** You can walk over the mountains to your campsite or take a safer, easier trip around the mountains. You are on the ground at $P(0, 0)$ and the campsite is at $C(-2, 7)$. You travel around the mountain by going to $A(2, 5)$. The coordinate system is measured in meters. Draw a diagram of the situation. Find PA and AC.

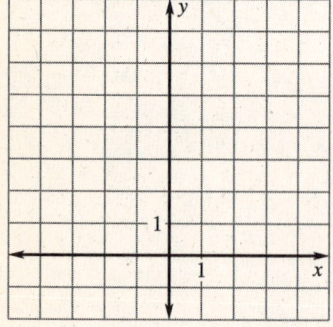

Answers

1. _____
2. _____
3. _____
4. _____
5. _____
6. _____
7. diagram: use grid at left
   _____

LESSON 1.4

Teacher's Name _____ Class _____ Room _____ Date _____

Lesson Plan
1-day lesson (See *Pacing the Chapter,* TE pages 1C–1D)   For use with pages 26–32

GOALS   1. Use angle postulates.
2. Classify angles as acute, right, obtuse, or straight.

State/Local Objectives _____
_____

✓ **Check the items you wish to use for this lesson.**

STARTING OPTIONS
____ Homework Check: TE page 21: Answer Transparencies
____ Warm-Up or Daily Homework Quiz: TE pages 26 and 24, CRB page 51, or Transparencies

TEACHING OPTIONS
____ Motivating the Lesson : TE page 27
____ Lesson Opener (Activity): CRB page 52 or Transparencies
____ Examples 1–4: SE pages 26–28
____ Extra Examples: TE pages 27–28 or Transparencies; Internet
____ Closure Question: TE page 28
____ Guided Practice Exercises: SE page 29

APPLY/HOMEWORK
Homework Assignment
____ Basic   17–25, 26–48 even, 51, 62–78 even
____ Average   17–25, 26–48 even, 51, 62–78 even
____ Advanced   17–25, 26–48 even, 51, 55–50, 62–78 even

Reteaching the Lesson
____ Practice Masters: CRB pages 53–55 (Level A, Level B, Level C)
____ Reteaching with Practice: CRB pages 56–57 or Practice Workbook with Examples
____ Personal Student Tutor

Extending the Lesson
____ Applications (Interdisciplinary): CRB page 59
____ Challenge: SE page 32; CRB page 60 or Internet

ASSESSMENT OPTIONS
____ Checkpoint Exercises: TE pages 27–28 or Transparencies
____ Daily Homework Quiz (1.4): TE page 32, CRB page 63, or Transparencies
____ Standardized Test Practice: SE page 31; TE page 32; STP Workbook; Transparencies

Notes _____
_____
_____

LESSON 1.4

TEACHER'S NAME _____ CLASS _____ ROOM _____ DATE _____

Lesson Plan for Block Scheduling
Half-day lesson (See *Pacing the Chapter,* TE pages 1C–1D) For use with pages 26–32

GOALS
1. Use angle postulates.
2. Classify angles as acute, right, obtuse, or straight.

State/Local Objectives _____

_____

_____

CHAPTER PACING GUIDE	
Day	Lesson
1	1.1 (all); 1.2 (all)
2	1.3 (all)
3	**1.4 (all)**; 1.5 (begin)
4	1.5 (end); 1.6 (begin)
5	1.6 (end); 1.7 (begin)
6	1.7 (end); Review Ch. 1
7	Assess Ch. 1; 2.1 (all)

✓ Check the items you wish to use for this lesson.

STARTING OPTIONS
____ Homework Check: TE page 21: Answer Transparencies
____ Warm-Up or Daily Homework Quiz: TE pages 26 and 24, CRB page 51, or Transparencies

TEACHING OPTIONS
____ Motivating the Lesson : TE page 27
____ Lesson Opener (Activity): CRB page 52 or Transparencies
____ Examples 1–4: SE pages 26–28
____ Extra Examples: TE pages 27–28 or Transparencies; Internet
____ Closure Question: TE page 28
____ Guided Practice Exercises: SE page 29

APPLY/HOMEWORK
Homework Assignment (See also the assignment for Lesson 1.5.)
____ Block Schedule: 17–25, 26–48 even, 51, 62–78 even

Reteaching the Lesson
____ Practice Masters: CRB pages 53–55 (Level A, Level B, Level C)
____ Reteaching with Practice: CRB pages 56–57 or Practice Workbook with Examples
____ Personal Student Tutor

Extending the Lesson
____ Applications (Interdisciplinary): CRB page 59
____ Challenge: SE page 32; CRB page 60 or Internet

ASSESSMENT OPTIONS
____ Checkpoint Exercises: TE pages 27–28 or Transparencies
____ Daily Homework Quiz (1.4): TE page 32, CRB page 63, or Transparencies
____ Standardized Test Practice: SE page 31; TE page 32; STP Workbook; Transparencies

Notes _____

_____

_____

LESSON 1.4 — WARM-UP EXERCISES

For use before Lesson 1.4, pages 26–32

Fill in the blanks.

1. The endpoint of $\overrightarrow{RQ}$ is _____.

2. Line segments with equal measures are _____.

3. A rule of geometry that is accepted without proof is a _____.

4. A pair of opposite rays form a _____.

DAILY HOMEWORK QUIZ

For use after Lesson 1.3, pages 17–25

Points A, D, F, and X are on a segment in this order. AD = 15, AF = 22, and AX = 30. Find each length.

1. DF
2. FX

Use the Distance Formula to decide whether PQ = QR.

3. $P(2, 4)$, $Q(4, 10)$, $R(0, 6)$

4. $P(-1, -2)$, $Q(2, 0)$, $R(4, 3)$

LESSON 1.4

NAME _____ DATE _____

Activity Lesson Opener

For use with pages 26–32

Available as a transparency

SET UP: Work with a partner.

YOU WILL NEED: • scissors

A pizza was made in the pan at the right. The pizza was then cut into eight pieces. The pieces were placed on a serving tray along with two pieces from another pizza, as shown below.

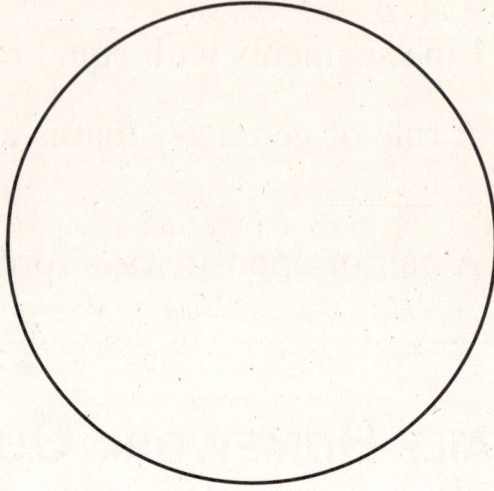

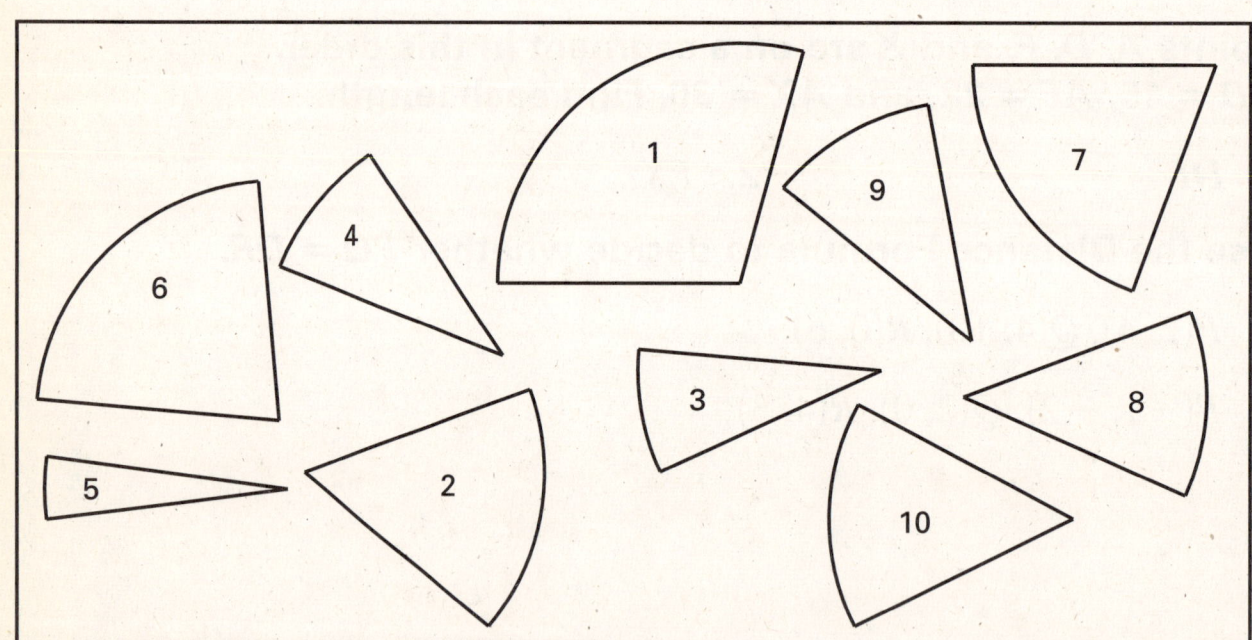

1. Which piece is the largest? Do you think that piece came from the pizza made in the pan shown? Why?

2. Cut out the pieces shown on the serving tray. Place the pieces in the circle to find the eight that came from the pan shown.

3. Explain how you know that the two pieces left over did not come from the pan shown.

LESSON 1.4

Practice A
For use with pages 26–32

Name the vertex and sides of the angle. Write two names for each angle.

1.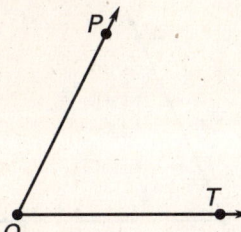
2.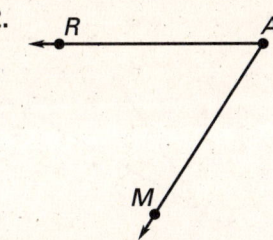
3.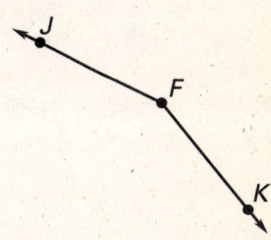

Use a protractor to measure each angle to the nearest degree.

4.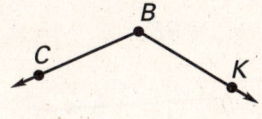
5.
6.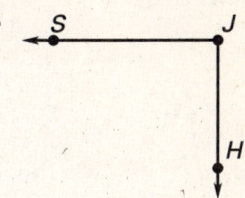

Use the Angle Addition Postulate to find the measure of the unknown angle.

7. $m\angle ABC = $ ___?___
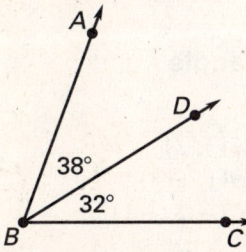

8. $m\angle DEF = $ ___?___

9. $m\angle MNR = $ ___?___
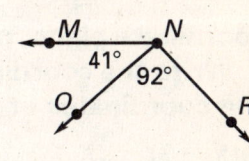

State whether the angle appears to be *acute*, *right*, *obtuse*, or *straight*. Then estimate its measure.

10.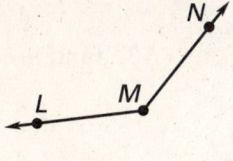
11.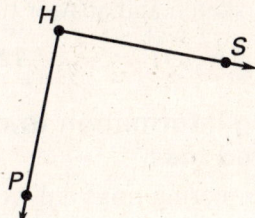
12.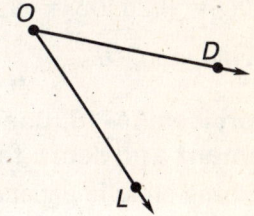

In a coordinate plane, plot the points and sketch $\angle ABC$. Classify the angle. Write the coordinates of a point that lies in the interior of the angle and the coordinates of a point that lies in the exterior of the angle.

13. $A(2, -4)$
    $B(-1, -1)$
    $C(4, 1)$

14. $A(-2, 1)$
    $B(1, 4)$
    $C(7, 2)$

15. $A(4, 3)$
    $B(2, -2)$
    $C(-3, 0)$

LESSON 1.4

Practice B
For use with pages 26–32

Use a protractor to measure each angle to the nearest degree. Write two names for each angle.

1.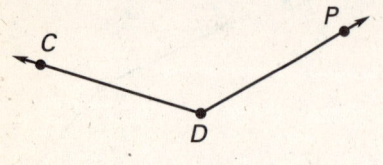
2.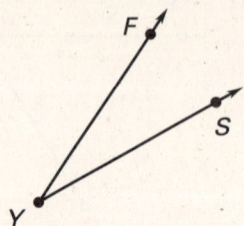
3.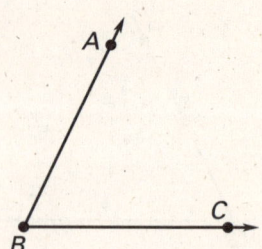

Use the Angle Addition Postulate to find the measure of the unknown angle.

4. $m\angle TMF = $?
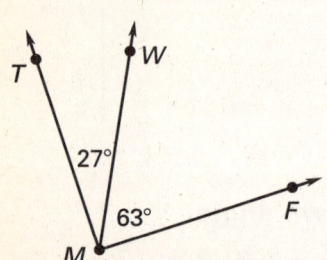

5. $m\angle XLV = $?
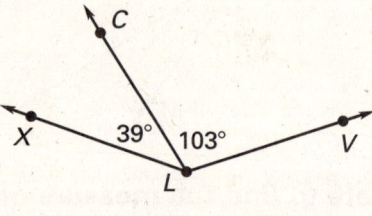

6. $m\angle WRF = $?
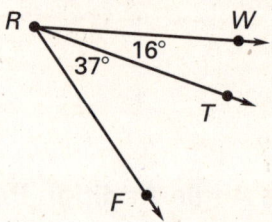

In a coordinate plane, plot the points and sketch $\angle ABC$. Classify the angle. Write the coordinates of a point that lies in the interior of the angle and the coordinates of a point that lies in the exterior of the angle.

7. $A(5, -3)$
   $B(-3, -1)$
   $C(2, 2)$

8. $A(-3, 0)$
   $B(1, 3)$
   $C(6, 0)$

9. $A(3, 2)$
   $B(1, -3)$
   $C(-4, -1)$

In Exercises 10–13, use the following information.

Q is in the interior of $\angle ROS$. S is in the interior of $\angle QOP$. P is in the interior of $\angle SOT$. S is in the interior of $\angle ROT$ and $m\angle ROT = 160°$. $m\angle SOT = 100°$ and $m\angle ROQ = m\angle QOS = m\angle POT$. Make a sketch and answer the following.

10. Find $m\angle QOP$.
11. Find $m\angle QOT$.
12. Find $m\angle ROQ$.
13. Find $m\angle SOP$.

In Exercises 14–18, use the following information to mark the placement and score for the indicated toss.

The scoring areas in a game are rings. The scoring rings are worth 100, 50, 25, and 10 points, as shown in the figure. For the ball that landed at point A, $m\angle BOA = 120°$ and $AO = 2.5$ in. The score for this ball is 50.

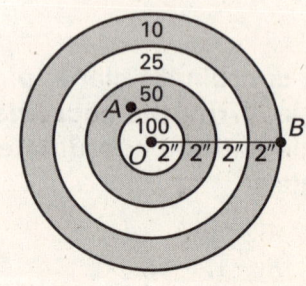

14. $AO = 3.5$ in., $m\angle BOA = 60°$
15. $AO = 1.4$ in., $m\angle BOA = 115°$
16. $AO = 4.5$ in., $m\angle BOA = 180°$
17. $AO = 5.5$ in., $m\angle BOA = 5°$
18. Find the total score for all four tosses.

LESSON 1.4

Practice C
For use with pages 26–32

Use a protractor to measure each angle to the nearest degree. Write two names for each angle.

1.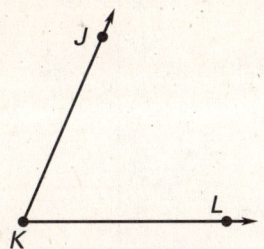
2.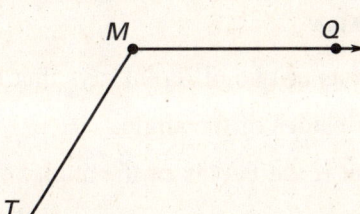
3.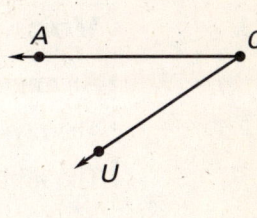

Use the Angle Addition Postulate to find the measure of the unknown angle.

4. $m\angle FDC = \underline{\quad ? \quad}$

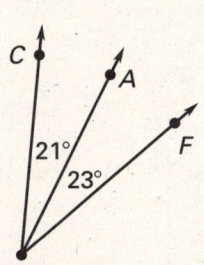

5. $m\angle CDE = \underline{\quad ? \quad}$

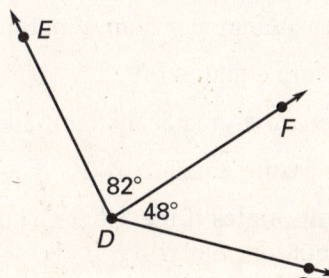

6. $m\angle XYZ = \underline{\quad ? \quad}$

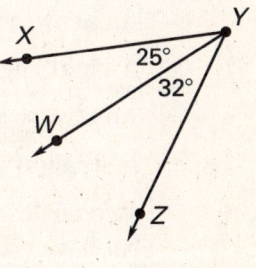

In a coordinate plane, plot the points and sketch $\angle ABC$. Classify the angle. Write the coordinates of a point that lies in the interior of the angle and the coordinates of a point that lies in the exterior of the angle.

7. $A(-5, -4)$
   $B(-3, 0)$
   $C(1, -4)$

8. $A(-5, 0)$
   $B(-1, -4)$
   $C(4, 2)$

9. $A(0, 1)$
   $B(-2, -4)$
   $C(-7, -2)$

In Exercises 10–13, use the following information.

Q is in the interior of $\angle ROS$. S is in the interior of $\angle QOP$. P is in the interior of $\angle SOT$. $m\angle ROT = 127°$, $m\angle SOT = 71°$, and $m\angle ROQ = m\angle QOS = m\angle POT$. Make a sketch and answer the following.

10. Find $m\angle QOP$
11. Find $m\angle QOT$
12. Find $m\angle ROQ$
13. Find $m\angle SOP$

Let Q be in the interior of $\angle POR$. Use the Angle Addition Postulate to solve for x. Find the measure of each angle.

14. $m\angle POQ = (x + 4)°$
    $m\angle QOR = (2x - 2)°$
    $m\angle POR = 26°$

15. $m\angle POQ = (3x + 7)°$
    $m\angle QOR = (5x - 2)°$
    $m\angle POR = 61°$

16. $m\angle POQ = \left(\frac{1}{3}x + \frac{1}{3}\right)°$
    $m\angle QOR = \left(2x + \frac{4}{3}\right)°$
    $m\angle POR = (5x - 1)°$

LESSON 1.4

Reteaching with Practice

For use with pages 26–32

GOAL Use angle postulates and classify angles as acute, right, obtuse, or straight

VOCABULARY

An **angle** consists of two different rays that have the same initial point.

The rays are the **sides** of the angle.

The initial point is the **vertex** of the angle.

Angles that have the same measure are called **congruent angles.**

A point is in the **interior** of an angle if it is between points that lie on each side of the angle.

A point is in the **exterior** of an angle if it is not on the angle or in its interior.

An **acute** angle has measure greater than 0° and less than 90°.

A **right** angle has measure equal to 90°.

An **obtuse** angle has measure greater than 90° and less than 180°.

A **straight** angle has measure equal to 180°.

Two angles are **adjacent angles** if they share a common vertex and side, but have no common interior points.

Postulate 3 Protractor Postulate:

Consider a point A on one side of $\overleftrightarrow{OB}$. The rays of the form $\overrightarrow{OA}$ can be matched one to one with the real numbers from 0 to 180.

The **measure** of $\angle AOB$ is equal to the absolute value of the difference between the real numbers for $\overrightarrow{OA}$ and $\overrightarrow{OB}$.

Postulate 4 Angle Addition Postulate:

If P is in the interior of $\angle RST$, then $m\angle RSP + m\angle PST = m\angle RST$.

EXAMPLE 1 Naming Angles

a. Write three names for the angle and name the vertex and sides of the angle.

b. Suppose R is in the interior of $\angle NOP$, with $m\angle NOR = 23°$ and $m\angle ROP = 27°$. Find $m\angle NOP$.

SOLUTION

a. $\angle NOP$, $\angle PON$, and $\angle O$ are all appropriate names for this angle. The vertex of this angle is point O and the sides are $\overrightarrow{ON}$ and $\overrightarrow{OP}$.

b. By Angle Addition Postulate, $m\angle NOP = m\angle NOR + m\angle ROP = 23° + 27° = 50°$

LESSON 1.4 CONTINUED

Reteaching with Practice

For use with pages 26–32

Exercises for Example 1

Write three names for the angles and name the vertex and sides of each.

1.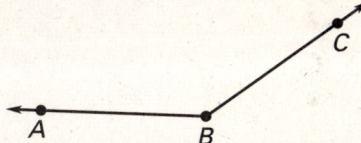

2.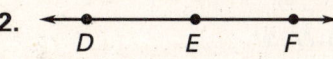

3. Suppose that the angle at the right measures 60° and that there is a point K in the interior of the angle such that $m\angle GHK = 25°$. Find $m\angle KHI$.

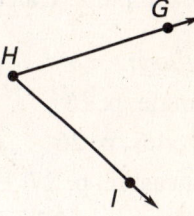

EXAMPLE 2 Classifying Angles in a Coordinate Plane

Plot the points $A(1, 1)$, $B(-1, 1)$, $C(1, 3)$, $D(3, 2)$, and $E(3, 1)$. Then classify the following angles as acute, right, obtuse, or straight.

a. $\angle CAB$ b. $\angle DAE$ c. $\angle BAD$ d. $\angle EAB$

SOLUTION

Begin by plotting the points, then observe whether each angle is less than 90°, equal to 90°, between 90° and 180°, or equal to 180°.

a. right angle   b. acute angle
c. obtuse angle d. straight angle

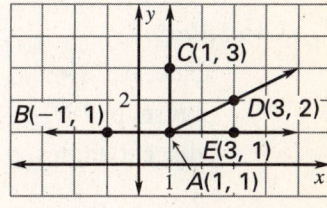

Exercises for Example 2

Plot the given points and classify the given angles as *acute*, *right*, *obtuse*, or *straight*.

4. $A(-2, 4)$, $B(-5, 1)$, $C(0, 0)$, and $D(3, 0)$
   a. $\angle ACB$
   b. $\angle BCD$
   c. $\angle ACD$

5. $E(4, 0)$, $F(3, 2)$, $G(1, 0)$, $H(-1, -2)$, $I(-1, 2)$
   a. $\angle HGF$
   b. $\angle EGF$
   c. $\angle EGI$
   d. $\angle FGI$
   e. $\angle HGI$

LESSON 1.4

NAME _____ DATE _____

Quick Catch-Up for Absent Students
For use with pages 26–32

The items checked below were covered in class on (date missed) _____

Lesson 1.4: Angles and Their Measures

____ **Goal 1:** Use angle postulates. (pp. 26–27)

Material Covered:

　　____ Example 1: Naming Angles

　　____ Student Help: Study Tip

　　____ Example 2: Calculating Angle Measures

Vocabulary:

　　angle, p. 26　　　　　　　　sides, p. 26
　　vertex, p. 26　　　　　　　 congruent angles, p. 26
　　measure, p. 27　　　　　　 interior, p. 27
　　exterior, p. 27

____ **Goal 2:** Classify angles as acute, right, obtuse, or straight. (p. 28)

Material Covered:

　　____ Student Help: Study Tip

　　____ Example 3: Classifying Angles in a Coordinate Plane

　　____ Example 4: Drawing Adjacent Angles

Vocabulary:

　　acute, p. 28　　　　　　　　right, p. 28
　　obtuse, p. 28　　　　　　　 straight, p. 28
　　adjacent angles, p. 28

____ Other (specify) _____
_____

Homework and Additional Learning Support

　　____ Textbook (specify) pp. 29–32 _____
_____

　　____ Internet: Extra Examples at www.mcdougallittell.com

　　____ *Reteaching with Practice* worksheet (specify exercises) _____

　　____ *Personal Student Tutor* for Lesson 1.4

Lesson 1.4

Interdisciplinary Application
For use with pages 26–32

Field of Vision

HEALTH In health class, you study eye care and eye disease. You learn that tunnel vision is a serious sight impairment that can reduce a person's field of vision, the physical space visible to the eye in a certain position. Looking directly forward, the eye can see about 60° inward (toward the nose) and about 90° outward (toward the side of the face). The outer range of vision is called peripheral vision.

Let $\angle OLS$ and $\angle PRT$ represent the outward vision angle of 90°. Let $\angle SLI$ and $\angle TRI$ represent the inward vision angle of 60°.

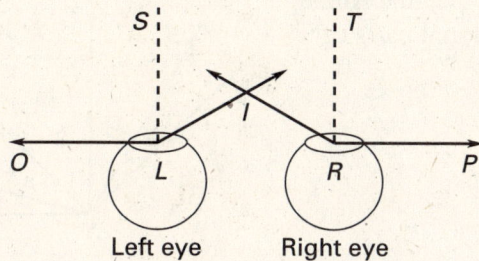

Left eye   Right eye

1. Name the angle that represents the total range of vision for the right eye.
2. Find the measure of the angle from Exercise 1.
3. If peripheral vision in each eye were reduced so that the outward vision angle was congruent to the inward vision angle, what would the total range of vision be in the right eye?
4. Severe tunnel vision is classified as legal blindness if the total range of vision in each eye is reduced to 20° or less. What would the measures of the inward and outward angles of vision be for each eye?
5. What percent of vision is lost with severe tunnel vision? How would that impair a person's ability to drive a car?

LESSON 1.4

Challenge: Skills and Applications
For use with pages 26–32

1. Suppose $\angle 1 \cong \angle 3$. Use the Angle Addition Postulate to show that $\angle BAD \cong \angle CAE$.

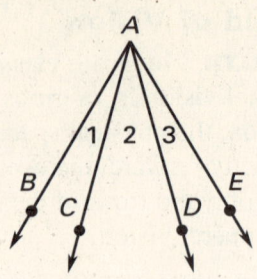

2. Suppose $\angle 2 \cong \angle 4$ and $\angle RXT \cong \angle UXW$. Use the Angle Addition Postulate to write as many additional pairs of congruent angles as possible.

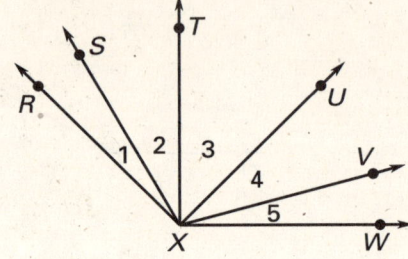

In Exercises 3–6, use the diagram shown.

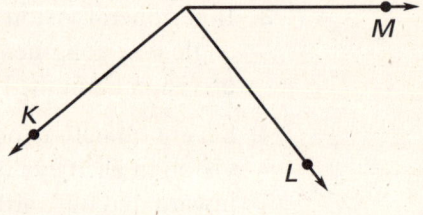

3. If $m\angle KJM = 145°$, $m\angle KJL = 6x°$, and $m\angle LJM = (3x + 10)°$, find x.

4. If $m\angle KJM = (2x^2 + 6x - 5)°$, $m\angle KJL = 85°$, and $m\angle LJM = (x^2 + 1)°$, find x. (*Hint:* Make sure your answer is reasonable!)

5. If $3(m\angle LJM) = 2(m\angle KJL)$ and $m\angle KJM = 140°$, find $m\angle KJL$.

6. If the ratio of $m\angle KJL$ to $m\angle LJM$ is 5:3, and $m\angle KJM = 144°$, find $m\angle LJM$.

In Exercises 7 and 8, consider the following statement.

If $\angle RSP$ and $\angle PST$ are adjacent angles, then $m\angle RSP + m\angle PST = m\angle RST$.

7. Explain how the statement differs from the Angle Addition Postulate.

8. Sketch a counterexample to show that the statement is false.

In Exercises 9–11, assume that $\angle RSP$ and $\angle PST$ are adjacent angles, and assume that $m\angle RSP = m\angle PST = x°$.

9. If $\angle RSP$ is acute, write an expression for $m\angle RST$ in terms of x.

10. If $\angle RSP$ is obtuse, write an expression for $m\angle RST$ in terms of x.

11. If $\angle RSP$ is a right angle, which of your two expressions gives $m\angle RST$ in terms of x? Explain.

LESSON 1.5

TEACHER'S NAME _____ CLASS _____ ROOM _____ DATE _____

Lesson Plan

2-day lesson (See *Pacing the Chapter,* TE pages 1C–1D) For use with pages 33–42

GOALS 1. Bisect a segment.
2. Bisect an angle.

State/Local Objectives _____
_____

✓ **Check the items you wish to use for this lesson.**

STARTING OPTIONS
____ Homework Check: TE page 29: Answer Transparencies
____ Warm-Up or Daily Homework Quiz: TE pages 34 and 32, CRB page 63, or Transparencies

TEACHING OPTIONS
____ Motivating the Lesson: TE page 35
____ Concept Activity: SE page 33
____ Lesson Opener (Application): CRB page 64 or Transparencies
____ Technology Activity with Keystrokes: CRB pages 65–68
____ Examples: Day 1: 1–4: SE pages 35–37; Day 2: 5, SE page 37
____ Extra Examples: Day 1: TE pages 35–37 or Transp.; Day 2: TE page 37 or Transp.
____ Closure Question: TE page 37
____ Guided Practice: SE page 38   Day 1: Exs. 1–13; Day 2: Exs. none

APPLY/HOMEWORK
Homework Assignment
____ Basic   Day 1: 14–42 even; Day 2: 44–54, 58, 62–72 even; Quiz 2: 1–6
____ Average   Day 1: 14–42 even; Day 2: 44–54, 56, 58, 62–72 even; Quiz 2: 1–6
____ Advanced   Day 1: 14–42 even; Day 2: 44–54, 56–60, 62–72 even; Quiz 2: 1–6

Reteaching the Lesson
____ Practice Masters: CRB pages 69–71 (Level A, Level B, Level C)
____ Reteaching with Practice: CRB pages 72–73 or Practice Workbook with Examples
____ Personal Student Tutor

Extending the Lesson
____ Applications (Real-Life): CRB page 75
____ Challenge: SE page 41; CRB page 76 or Internet

ASSESSMENT OPTIONS
____ Checkpoint Exercises: Day 1: TE pages 35, 37 or Transp.; Day 2: TE page 37 or Transp.
____ Daily Homework Quiz (1.5): TE page 41, CRB page 80, or Transparencies
____ Standardized Test Practice: SE page 41; TE page 41; STP Workbook; Transparencies
____ Quiz (1.4–1.5): SE page 42; CRB page 77

Notes _____
_____
_____

LESSON 1.5

Teacher's Name _____ **Class** _____ **Room** _____ **Date** _____

Lesson Plan for Block Scheduling

1-day lesson (See *Pacing the Chapter*, TE pages 1C–1D) For use with pages 33–42

GOALS
1. Bisect a segment.
2. Bisect an angle.

State/Local Objectives _____

_____

_____

CHAPTER PACING GUIDE	
Day	Lesson
1	1.1 (all); 1.2 (all)
2	1.3 (all)
3	1.4 (all); **1.5 (begin)**
4	**1.5 (end)**; 1.6 (begin)
5	1.6 (end); 1.7 (begin)
6	1.7 (end); Review Ch. 1
7	Assess Ch. 1; 2.1 (all)

✓ **Check the items you wish to use for this lesson.**

STARTING OPTIONS
____ Homework Check: TE page 29: Answer Transparencies
____ Warm-Up or Daily Homework Quiz: TE pages 34 and 32, CRB page 63, or Transparencies

TEACHING OPTIONS
____ Motivating the Lesson: TE page 35
____ Concept Activity: SE page 33
____ Lesson Opener (Application): CRB page 64 or Transparencies
____ Technology Activity with Keystrokes: CRB pages 65–68
____ Examples: Day 3: 1–4: SE pages 35–37; Day 4: 5, SE page 37
____ Extra Examples: Day 3: TE pages 35–37 or Transp.; Day 4: TE page 37 or Transp.
____ Closure Question: TE page 37
____ Guided Practice: SE page 38 Day 3: Exs. 1–13; Day 4: Exs. none

APPLY/HOMEWORK
Homework Assignment (See also the assignments for Lessons 1.4 and 1.6.)
____ Block Schedule: Day 3: 14–42 even; Day 4: 44–54, 56, 58, 62–72 even; Quiz 2: 1–6

Reteaching the Lesson
____ Practice Masters: CRB pages 69–71 (Level A, Level B, Level C)
____ Reteaching with Practice: CRB pages 72–73 or Practice Workbook with Examples
____ Personal Student Tutor

Extending the Lesson
____ Applications (Real-Life): CRB page 75
____ Challenge: SE page 41; CRB page 76 or Internet

ASSESSMENT OPTIONS
____ Checkpoint Exercises: Day 3: TE pages 35, 37 or Transp.; Day 4: TE page 37 or Transp.
____ Daily Homework Quiz (1.5): TE page 41, CRB page 80, or Transparencies
____ Standardized Test Practice: SE page 41; TE page 41; STP Workbook; Transparencies
____ Quiz (1.4–1.5): SE page 42; CRB page 77

Notes _____

_____

_____

LESSON 1.5

WARM-UP EXERCISES
For use before Lesson 1.5, pages 33–42

Solve each equation.

1. $\dfrac{x+1}{2} = 3$

2. $\dfrac{y-3}{-1} = 4$

3. $x + 30 = 2x - 20$

4. $3y + 15 = 5y - 20$

DAILY HOMEWORK QUIZ
For use after Lesson 1.4, pages 26–32

In a coordinate plane, plot the points $A(-6, 2)$, $B(0, 0)$, and $C(1, 3)$ and sketch $\angle ABC$. Then answer the questions below.

1. Name the vertex.

2. Write two names for the angle.

3. Find $m\angle ABC$.

4. Write the coordinates of a point that lies in the interior of the angle.

LESSON 1.5

Application Lesson Opener
For use with pages 34–42

A diagram of a park is shown below.

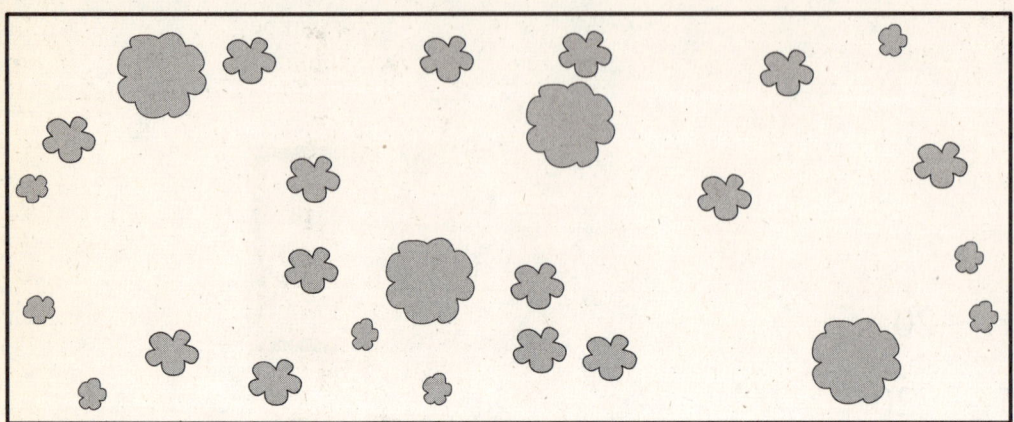

1. Members of the city council want to build a sidewalk through the middle of the park. They decide this sidewalk should not go through the park diagonally. What do you think it means to build a sidewalk through the middle of the park? Draw a sidewalk on the diagram above that meets these conditions.

2. After the first sidewalk is built, they decide to build four more sidewalks, one at each of the four corners of the park. They want each corner sidewalk to divide the angle at the corner in half. Draw a sidewalk at each corner in the diagram above that meets these conditions.

3. Do any of the sidewalks you drew in Questions 1 and 2 intersect?

4. Compare your diagram with the diagrams your classmates drew. How are your diagrams the same? How are they different?

LESSON 1.5

Technology Activity

For use with pages 34–42

GOAL To learn the basics of geometry software

In this activity, the basics of geometry are explored. Geometry software can be used to construct and measure geometric figures and to calculate and tabulate data.

Activity

① Open a geometry file.

② Construct a triangle.

③ In the triangle, measure the length of the three sides and measure the three angles.

④ Calculate the sum of the angles of the triangle.

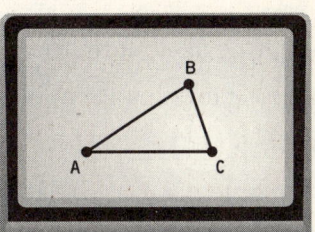

Exercises

1. What is the sum of the angles in a triangle?

2. If the measures of two angles in a triangle are 45° and 75°, what is the measure of the third angle?

LESSON 1.5 CONTINUED

Technology Activity Keystrokes
For use with pages 34–42

TI-92

1. Open a geometry file **APPS** 8 3 Cursor to Variable and enter a name of 8 characters or less.

   ENTER **ENTER**

2. Construct sides *AB*, *BC*, *AC* to form triangle *ABC*

   F2 5 (Place cursor at location for point *A*) **ENTER** *A* (Move cursor to location for point *B*) **ENTER** *B* **ENTER** (Move cursor to location for point *C*) **ENTER** *C* **ENTER** (Move cursor to point *A*) **ENTER**

3. Measure the lengths of the sides of triangle *ABC*

   F6 1 (Move cursor to side *AB*) **ENTER** (Move cursor to side *BC*) **ENTER** (Move cursor to side *AC*) **ENTER**

   Measure the angles of triangle *ABC*

   F6 3 (Move cursor to point *C*) **ENTER** (Move cursor to point *A*) **ENTER** (Move cursor to point *B*) **ENTER** (Note: this is the measure of angle *CAB*) (Move cursor to point *B*) **ENTER** (Move cursor to point *C*) **ENTER** (Move cursor to point *A*) **ENTER** (Note: this is the measure of angle *BCA*) (Move cursor to point *A*) **ENTER** (Move cursor to point *B*) **ENTER** (Move cursor to point *C*) **ENTER** (Note: this is the measure of angle *ABC*)

4. Calculate the sum of the angle measures

   F6 6 (Press the cursor key until an angle measure is highlighted) **ENTER** **+** (Press the cursor key until a different angle measure is highlighted) **ENTER** **+** (Press the cursor key until the last angle measure is highlighted) **ENTER**

LESSON 1.5 CONTINUED

Technology Activity Keystrokes

For use with pages 34–42

SKETCHPAD

1. Open Sketchpad.

2. Construct sides *AB*, *BC*, and *AC* to form triangle *ABC*
   Choose the segment straightedge tool, then draw each side to form triangle *ABC*.

3. Measure the length of the sides of triangle *ABC*
   Choose the translate selection arrow tool and select side *AB*. Then hold down the shift key and select sides *BC* and *AC*. Choose **Length** from the **Measure** menu.

   Measure the angles of triangle *ABC*
   To measure angle *ABC*, select point *A*, hold down the shift key and select points *B* and *C* (in that order) and choose **Angle** from the **Measure** menu. To measure angle *CAB*, select point *C*, hold down the shift key and select points *A* and *B* (in that order), and choose **Angle** from the **Measure** menu. To measure angle *BCA*, select point *B*, hold down the shift key and select points *C* and *A* (in that order), and choose **Angle** from the **Measure** menu.

4. To calculate the sum of the angles of triangle *ABC*, choose **Calculate** from the **Measure** menu. Then click on the measure of angle *ABC*, [+], click on the measure of angle *CAB*, [+], click on the measure of angle *BCA*, and click OK.

LESSON 1.5

Technology Activity Keystrokes

For use with page 39

Keystrokes for Exercise 43

TI-92

1. Draw a triangle

   F3 3 (Place cursor to desired location of first vertex) **ENTER** (Move cursor to desired location of second vertex) **ENTER** (Move cursor to desired location of third vertex) **ENTER**

2. Construct an angle bisector of one of the angles of the triangle

   F4 5 (Move cursor to a side of desired angle) **ENTER** (Move cursor to vertex of desired angle) **ENTER** (Move cursor to other side of desired angle) **ENTER**

3. Find the midpoint of the side opposite of the bisected angle

   F4 3 (Move cursor to side opposite of the bisected angle) **ENTER**

4. Change the triangle

   F1 1 (Move cursor to a vertex of the triangle)

Use the drag key and the cursor pad to drag the vertex and change the triangle.

SKETCHPAD

1. Draw a triangle. Choose the segment straightedge tool and draw triangle *ABC*. (Hold down the mouse to begin the segment. Release the mouse to finish the segment.)

2. Construct an angle bisector of one of the angles of the triangle. Choose the selection arrow tool. Select vertex *C*. Hold down the shift key and select vertices *A* and *B*. Choose **Angle Bisector** from the **Construct** menu.

3. Find the midpoint of the opposite side of the triangle. Select segment *BC* and choose **Point at Midpoint** from the **Construct** menu.

4. Change the triangle. Drag one of the vertices of the triangle.

LESSON 1.5

NAME _____ DATE _____

Practice A
For use with pages 34–42

Use a ruler to measure and redraw the line segment on a piece of paper. Then use construction tools to find the segment bisector.

1. •———————•
   A   B

2. •————————————————•
   T   W

3. •——————————————•
   K   N

Find the coordinates of the midpoint of $\overline{AB}$.

4. $A(-2, 4), B(4, 0)$

5. $A(-3, -2), B(5, 4)$

6. $A(3, 1), B(-5, -4)$

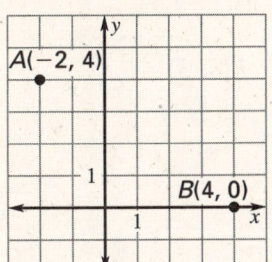
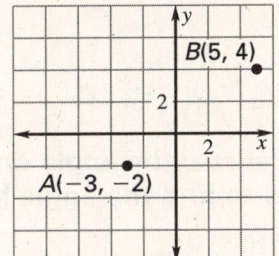
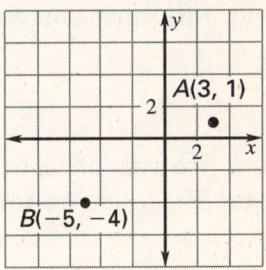

Find the coordinates of the midpoint of a segment with the given endpoints.

7. $A(0, 0)$
   $B(4, 6)$

8. $C(2, 4)$
   $D(0, -8)$

9. $E(-3, 2)$
   $F(7, -6)$

10. $G(5, -6)$
    $H(3, -3)$

Use a protractor to measure and redraw the angle on a piece of paper. Then use construction tools to find the angle bisector.

11.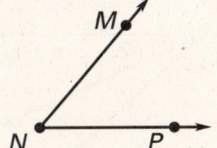

12.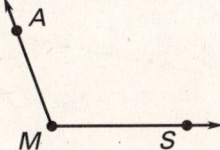

13.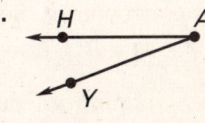

$\overrightarrow{PT}$ is the angle bisector of $\angle RPS$. Find the two angle measures not given in the diagram.

14.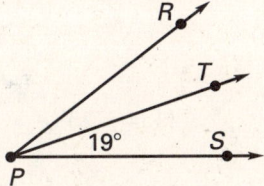

15.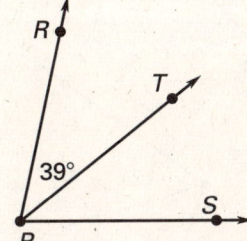

16.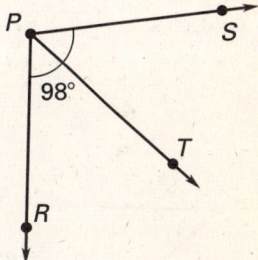

LESSON 1.5

Practice B
For use with pages 34–42

Use a ruler to measure and redraw the line segment on a piece of paper. Then use construction tools to find the segment bisector.

1. A———————B

2. T————————————W

Find the coordinates of the midpoint of a segment with the given endpoints.

3. $A(-3, 5)$
   $B(5, -1)$

4. $C(-4, -3)$
   $D(6, 3)$

5. $E(5, 0)$
   $F(-3, -5)$

Find the coordinates of the other endpoint of the segment with the given endpoint and midpoint M.

6. $T(6, 2)$
   $M(2, 0)$

7. $A(-4, 3)$
   $M(-1, -1)$

8. $P(7, 3)$
   $M(2, 1)$

Use a protractor to measure and redraw the angle on a piece of paper. Then use construction tools to find the angle bisector.

9.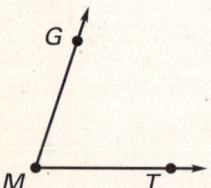

10.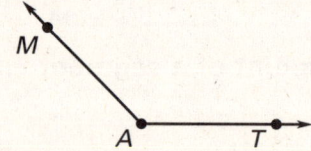

11.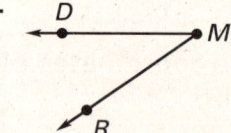

$\overrightarrow{PT}$ **is the angle bisector of** $\angle RPS$. **Find the two angle measures not given in the diagram.**

12.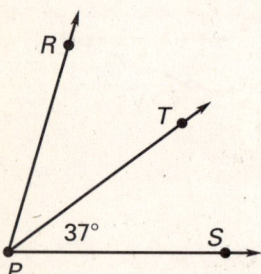

13.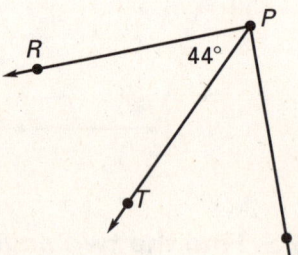

14.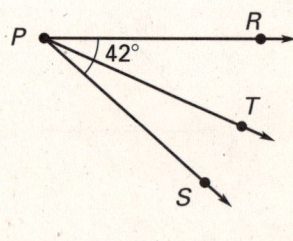

$\overrightarrow{BT}$ **bisects** $\angle ABC$. **Find the value of x.**

15.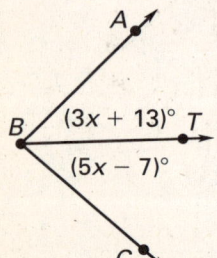

16.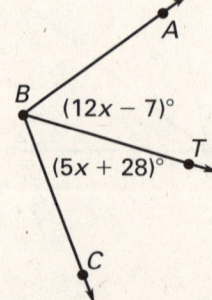

17.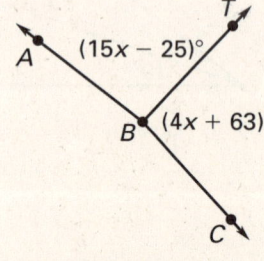

LESSON 1.5

NAME _____ **DATE** _____

Practice C
For use with pages 34–42

Use a ruler to measure and redraw the line segment on a piece of paper. Then use construction tools to find the segment bisector.

1.

2.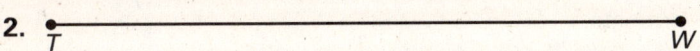

Find the coordinates of the midpoint of a segment with the given endpoints.

3. $A(-7, 2)$
   $B(3, 0)$

4. $G(3.2, 7.8)$
   $H(-2, 5)$

5. $K(-3.5, 2.7)$
   $L(7.3, -6.1)$

Find the coordinates of the other endpoint of the segment with the given endpoint and midpoint M.

6. $T(4, 1)$
   $M(3, 0)$

7. $A(-4.4, 3)$
   $M(-1.6, -1)$

8. $P(7.8, 3.5)$
   $M(2, -1.2)$

Use a protractor to measure and redraw the angle on a piece of paper. Then use construction tools to find the angle bisector.

9.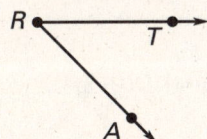

10.

11.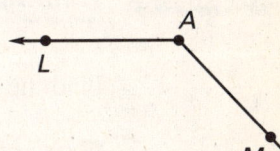

$\overrightarrow{PT}$ **is the angle bisector of** $\angle RPS$**. Find the two angle measures not given in the diagram.**

12.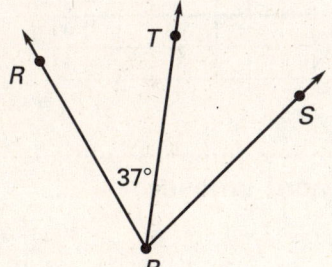

13.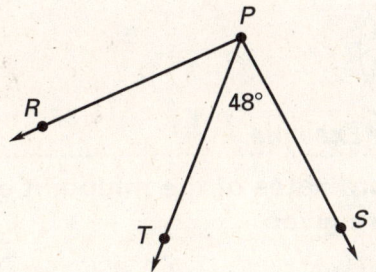

14.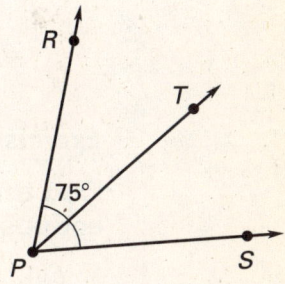

$\overrightarrow{BT}$ **bisects** $\angle ABC$**. Find the value of x.**

15.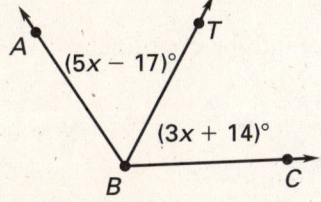

16.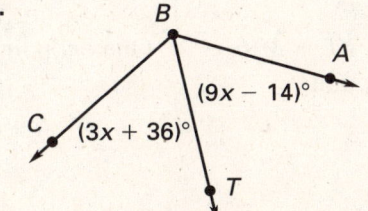

17.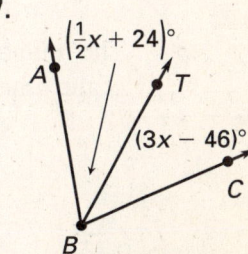

LESSON 1.5

Reteaching with Practice

For use with pages 34–42

GOAL Bisect a segment and bisect an angle

VOCABULARY

The **midpoint** of a segment is the point that divides, or **bisects**, the segment into two congruent segments.

A **segment bisector** is a segment, ray, line, or plane that intersects a segment at its midpoint.

A **construction** is a geometric drawing that uses a limited set of tools, usually a **compass** and a **straightedge**.

An **angle bisector** is a ray that divides an angle into two adjacent angles that are congruent.

The Midpoint Formula:

If $A(x_1, y_1)$ and $B(x_2, y_2)$ are points in a coordinate plane, then the midpoint of $\overline{AB}$ has coordinates $\left(\dfrac{x_1 + x_2}{2}, \dfrac{y_1 + y_2}{2}\right)$.

EXAMPLE 1 Finding the Coordinates of the Midpoint of a Segment

Find the coordinates of the midpoint of $\overline{CD}$ with endpoints $C(4, 3)$ and $D(-2, 0)$.

SOLUTION

Use the Midpoint Formula as follows.

$$M = \left(\dfrac{4 + (-2)}{2}, \dfrac{3 + 0}{2}\right)$$

$$= \left(1, \dfrac{3}{2}\right)$$

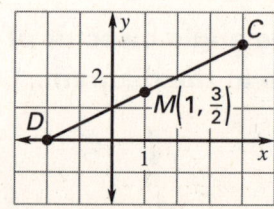

Exercises for Example 1

Find the coordinates of the midpoint of the segment whose endpoints are given.

1. $E(4, -4), F(1, 7)$
2. $G(2, 9), H(-3, 6)$
3. $I(-8, 3), J(3, 0)$

EXAMPLE 2 Finding the Coordinates of the Endpoint of a Segment

The midpoint of $\overline{KL}$ is $M(6, -2)$. One endpoint is $K(4, 3)$. Find the coordinates of the other endpoint.

LESSON 1.5 CONTINUED

Reteaching with Practice

For use with pages 34–42

SOLUTION

Let (x, y) be the coordinates of L. Use the Midpoint Formula to write equations involving x and y.

$$\frac{4 + x}{2} = 6 \qquad \frac{3 + y}{2} = -2$$
$$4 + x = 12 \qquad 3 + y = -4$$
$$x = 8 \qquad y = -7$$

So, the other endpoint of the segment is $L(8, -7)$.

Exercises for Example 2

Find the coordinates of the other endpoint of a segment with the given endpoint and midpoint M.

4. $N(-1, 5), M(0, 1)$
5. $P(6, -4), M(3, 10)$
6. $R(-7, -3), M(0, 0)$

EXAMPLE 3 Finding the Measure of an Angle

In the diagram, $\vec{BC}$ bisects $\angle ABD$. Solve for x.

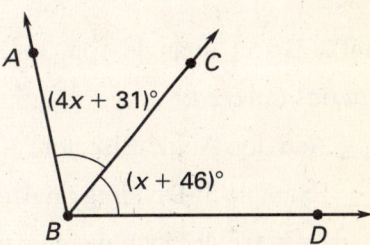

SOLUTION

$m\angle ABC = m\angle CBD$ Congruent angles have equal measures.
$(4x + 31)° = (x + 46)°$ Substitute given measures.
$4x = x + 15$ Subtract 31° from each side.
$3x = 15$ Subtract x from each side.
$x = 5$ Divide each side by 3.

Exercises for Example 3

$\vec{BD}$ **bisects** $\angle ABC$. **Find the value of x.**

7.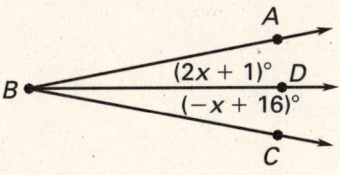

8.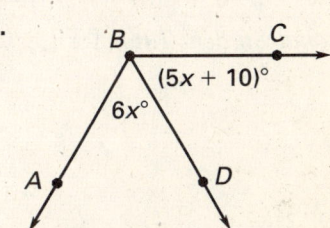

LESSON 1.5

Quick Catch-Up for Absent Students
For use with pages 33–42

NAME _____ DATE _____

The items checked below were covered in class on (date missed) _____

Activity 1.5: Folding Bisectors (p. 33)

____ **Goal:** Divide a Segment or an angle into two equal parts.

Lesson 1.5: Segment and Angle Bisectors

____ **Goal 1:** Bisect a segment. (pp. 34–35)

Material Covered:

    ____ Activity: Segment Bisector and Midpoint

    ____ Example 1: Finding the Coordinates of the Midpoint of a Segment

    ____ Student Help: Study Tip

    ____ Example 2: Finding the Coordinates of the Endpoint of a Segment

Vocabulary:

    midpoint, p. 34   bisect, p. 34
    segment bisector, p. 34   compass, p. 34
    straightedge, p. 34   construct, p. 34
    construction, p. 34   Midpoint Formula, p. 35

____ **Goal 2:** Bisect an angle. (pp. 36–37)

Material Covered:

    ____ Activity: Angle Bisector

    ____ Example 3: Dividing an Angle Measure in Half

    ____ Example 4: Doubling an Angle Measure

    ____ Example 5: Finding the Measure of an Angle

Vocabulary:

    angle bisector, p. 36

____ Other (specify) _____

_____

Homework and Additional Learning Support

    ____ Textbook (specify) <u>pp. 38–42</u> _____

_____

    ____ *Reteaching with Practice* worksheet (specify exercises) _____

    ____ *Personal Student Tutor* for Lesson 1.5

Lesson 1.5

Real-Life Application: When Will I Ever Use This?

For use with pages 34–42

Billiards

When playing billiards, one of the most difficult skills to master is the bank shot. In this shot, you do not shoot directly at a pocket, but rather you ricochet the ball off of a cushion, or a side of the table, so that it goes into a pocket. Using angle bisectors, you can determine the target point on the cushion. In the diagram below, there is a ball that you wish to bank into the side pocket. In your mind, draw a perpendicular line from the ball to the bank. Next draw a line to the side pocket opposite the target pocket. If you can find the bisector of the angle for which the ball is the vertex, follow that bisector to the cushion. That is the spot to hit. The ball will bounce off of the cushion at the same angle that it approached it and ultimately go into the side pocket.

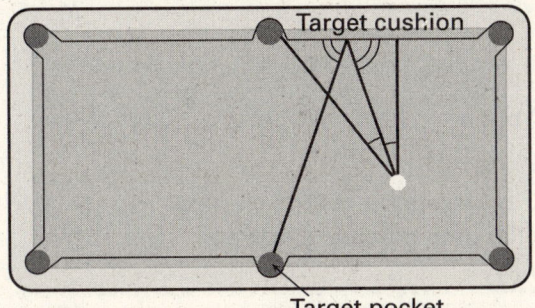

In Exercises 1–4, use the information above.

1. Show how to bank the gray ball in the diagram at the right off of target cushion 1 into target pocket 1. Bisect the angle by sight.

2. Bisect the angle again by using a compass. Was your guess close? Draw the path of the ball using a ruler and estimating the angle of reflection to see if you would make the shot.

3. Repeat the process in Exercises 1 and 2 to the same ball off of target cushion 2 into target pocket 2.

4. Is it possible to make the shot in Exercise 3? Explain.

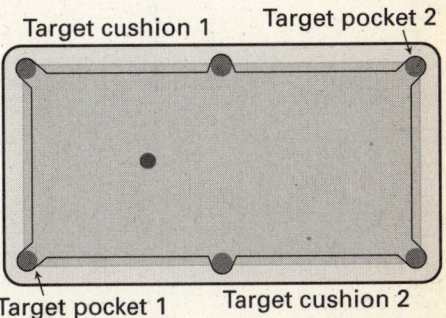

Lesson 1.5 Challenge: Skills and Applications

For use with pages 34–42

In Exercises 1–4, C is the midpoint of both $\overline{AE}$ and $\overline{BD}$.

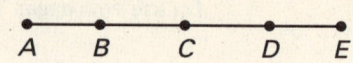

1. If $BC = x^2 - 18$ and $CD = x + 2$, find x.
2. If $AC = 2x - 1$ and $AE = x^2 - 2$, find x.
3. Can you be certain that $AB = DE$? Explain.
4. If $AB = 2x + 3$ and $DE = x^2$, what are the possible values of x?
5. Let $\angle PQR$ be an angle, and let M be the midpoint of $\overline{PR}$.

    Can you conclude that $\overrightarrow{QM}$ bisects $\angle PQR$? If so, explain why. If not, sketch a counterexample.

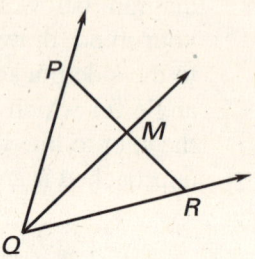

6. Suppose $\overrightarrow{AC}$ bisects $\angle BAD$, $\overrightarrow{AD}$ bisects $\angle BAE$, $\overrightarrow{AE}$ bisects $\angle BAF$. What is the maximum possible measure of $\angle BAC$?

7. Suppose $\overrightarrow{PK}$ bisects $\angle JPL$ and $\overrightarrow{PL}$ bisects $\angle KPM$. If $m\angle JPM = 150°$, find *both* possible measures of $\angle JPK$. Sketch both possible situations.

8. Suppose $\angle AXB \cong \angle BXC \cong \angle CXD \cong \angle DXE \cong \angle EXA$, and $\overrightarrow{XD}$ bisects $\angle AXB$. What is $m\angle AXB$? Sketch this situation.

In Exercises 9–14, use the following information to find the midpoint M of $\overline{PQ}$.

If $A(x_1, y_1, z_1)$ and $B(x_2, y_2, z_2)$ are two points in a three-dimensional coordinate system, then the midpoint of $\overline{AB}$ has coordinates $\left(\dfrac{x_2 + x_1}{2}, \dfrac{y_2 + y_1}{2}, \dfrac{z_2 + z_1}{2} \right)$.

9. $P(5, 7, -5)$
    $Q(3, 7, 11)$

10. $P(2, 0, 8)$
    $Q(2, 6, -2)$

11. $P(-3, 2, 7)$
    $Q(-11, 6, 3)$

12. $P(2, 0, 7)$
    $Q(8, -5, 12)$

13. $P(3.4, 1.8, 3.9)$
    $Q(6.2, -4.6, 0.3)$

14. $P(12, 35, 8.7)$
    $Q(1.6, -2.4, 1.9)$

LESSON 1.5

NAME _____ **DATE** _____

Quiz 2
For use after Lessons 1.4 and 1.5

1. State the Angle Addition Postulate for the three angles shown at the right. *(Lesson 1.4)*

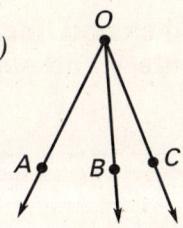

Answers

1. _____

2. diagram: use grid to left

3. diagram: use grid to left

4. diagram: use grid to left

5. diagram: use grid to left

6. _____

7. _____

In a coordinate plane, plot the points and sketch ∠RST. Classify the angle. Write the coordinates of a point that lies in the interior of the angle and the coordinates of a point that lies in the exterior of the angle. *(Lesson 1.4)*

2. $R(3, 1)$
   $S(0, 0)$
   $T(-3, 2)$

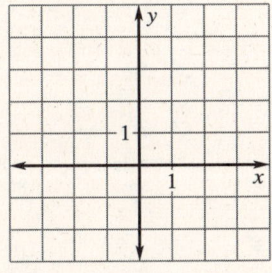

3. $R(0, 3)$
   $S(0, 0)$
   $T(4, 0)$

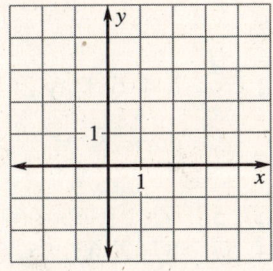

4. $R(-2, -2)$
   $S(-1, 3)$
   $T(1, -2)$

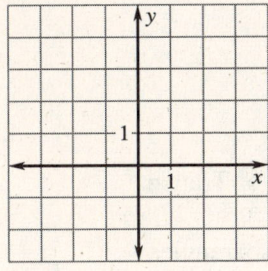

5. $R(-3, -3)$
   $S(0, 1)$
   $T(4, 0)$

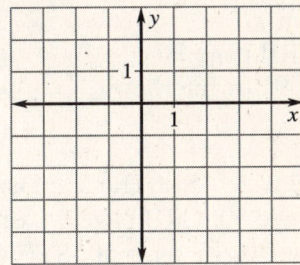

6. Find the midpoint of segment ST in Exercise 4.

7. In the diagram, $\vec{XS}$ is the angle bisector of ∠RXT. Find $m\angle SXT$ and $m\angle RXT$.

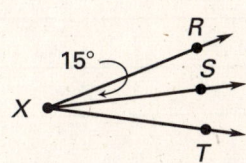

LESSON 1.6

Teacher's Name _____ Class _____ Room _____ Date _____

Lesson Plan
2-day lesson (See *Pacing the Chapter,* TE pages 1C–1D)

For use with pages 43–50

GOALS
1. Identify vertical angles and linear pairs.
2. Identify complementary and supplementary angles.

State/Local Objectives _____

_____

✓ **Check the items you wish to use for this lesson.**

STARTING OPTIONS
____ Homework Check: TE page 38: Answer Transparencies
____ Warm-Up or Daily Homework Quiz: TE pages 44 and 41, CRB page 80, or Transparencies

TEACHING OPTIONS
____ Lesson Opener (Activity): CRB page 81 or Transparencies
____ Technology Activity with Keystrokes: CRB pages 82–83
____ Examples: Day 1: 1–3: SE pages 44–45; Day 2: 4–6, SE page 46
____ Extra Examples: Day 1: TE page 45 or Transp.; Day 2: TE page 46 or Transp.; Internet
____ Technology Activity: SE page 43
____ Closure Question: TE page 46
____ Guided Practice: SE page 47   Day 1: Exs. 2, 4–6; Day 2: Exs. 1, 3, 7

APPLY/HOMEWORK
Homework Assignment
____ Basic   Day 1: 8–36; Day 2: 37–52, 57, 58, 62, 63, 65, 66–74 even
____ Average   Day 1: 8–36; Day 2: 37–54, 57, 58, 62, 63, 65, 66–74 even
____ Advanced   Day 1: 8–36; Day 2: 37–59, 62, 63, 65, 66–74 even

Reteaching the Lesson
____ Practice Masters: CRB pages 84–86 (Level A, Level B, Level C)
____ Reteaching with Practice: CRB pages 87–88 or Practice Workbook with Examples
____ Personal Student Tutor

Extending the Lesson
____ Applications (Interdisciplinary): CRB page 90
____ Challenge: SE page 50; CRB page 91 or Internet

ASSESSMENT OPTIONS
____ Checkpoint Exercises: Day 1: TE page 45 or Transp.; Day 2: TE page 46 or Transp.
____ Daily Homework Quiz (1.6): TE page 50, CRB page 94, or Transparencies
____ Standardized Test Practice: SE page 50; TE page 50; STP Workbook; Transparencies

Notes _____

_____

LESSON 1.6

TEACHER'S NAME _____ CLASS _____ ROOM _____ DATE _____

Lesson Plan for Block Scheduling
1-day lesson (See *Pacing the Chapter*, TE pages 1C–1D)   For use with pages 43–50

GOALS
1. Identify vertical angles and linear pairs.
2. Identify complementary and supplementary angles.

State/Local Objectives _____

_____

_____

CHAPTER PACING GUIDE	
Day	Lesson
1	1.1 (all); 1.2 (all)
2	1.3 (all)
3	1.4 (all); 1.5 (begin)
4	1.5 (end); **1.6 (begin)**
5	**1.6 (end)**; 1.7 (begin)
6	1.7 (end); Review Ch. 1
7	Assess Ch. 1; 2.1 (all)

✓ **Check the items you wish to use for this lesson.**

STARTING OPTIONS
____ Homework Check: TE page 38: Answer Transparencies
____ Warm-Up or Daily Homework Quiz: TE pages 44 and 41, CRB page 80, or Transparencies

TEACHING OPTIONS
____ Lesson Opener (Activity): CRB page 81 or Transparencies
____ Technology Activity with Keystrokes: CRB pages 82–83
____ Examples: Day 4: 1–3: SE pages 44–45; Day 5: 4–6, SE page 46
____ Extra Examples: Day 4: TE page 45 or Transp.; Day 5: TE page 46 or Transp.; Internet
____ Technology Activity: SE page 43
____ Closure Question: TE page 46
____ Guided Practice: SE page 47   Day 4: Exs. 2, 4–6; Day 5: Exs. 1, 3, 7

APPLY/HOMEWORK
Homework Assignment (See also the assignments for Lessons 1.5 and 1.7.)
____ Block Schedule: Day 4: 8–36; Day 5: 37–54, 57, 58, 62, 63, 65, 66–74 even

Reteaching the Lesson
____ Practice Masters: CRB pages 84–86 (Level A, Level B, Level C)
____ Reteaching with Practice: CRB pages 87–88 or Practice Workbook with Examples
____ Personal Student Tutor

Extending the Lesson
____ Applications (Interdisciplinary): CRB page 90
____ Challenge: SE page 50; CRB page 91 or Internet

ASSESSMENT OPTIONS
____ Checkpoint Exercises: Day 4: TE page 45 or Transp.; Day 5: TE page 46 or Transp.
____ Daily Homework Quiz (1.6): TE page 50, CRB page 94, or Transparencies
____ Standardized Test Practice: SE page 50; TE page 50; STP Workbook; Transparencies

Notes _____

_____

_____

LESSON 1.6

WARM-UP EXERCISES

For use before Lesson 1.6, pages 43–50

Name an example of each type of angle from the figure below.

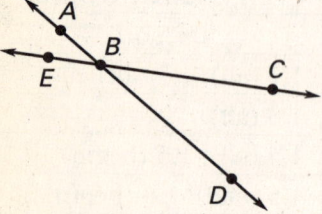

1. obtuse
2. acute
3. straight
4. a pair of adjacent angles

DAILY HOMEWORK QUIZ

For use after Lesson 1.5, pages 33–42

Find the coordinates of the midpoint of a segment with the given endpoints.

1. $A(-4, 6)$, $B(-2, 12)$
2. $C(0, 5)$, $D(-4, -5)$

Find the coordinates of the other endpoint of a segment with the given endpoint and midpoint M.

3. $L(-1, 9)$, $M(3, 7)$

$\vec{BD}$ bisects $\angle ABC$. Find the value of x.

4.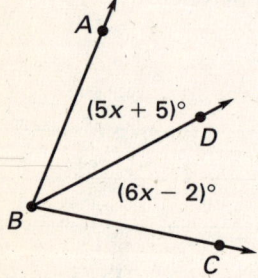

LESSON 1.6

Activity Lesson Opener
For use with pages 44–50

Part of a fence is shown below.

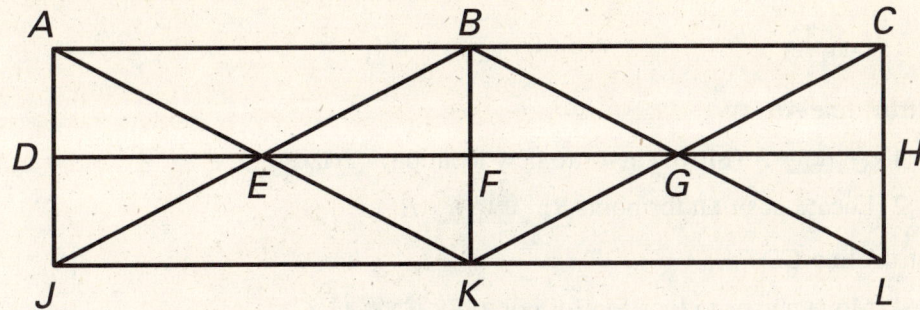

1. Choose two lines that intersect at point E, point F, or point G. Name the four angles formed by these two lines. Describe any relationships you see between any pair of angles formed by these two lines.

2. Choose two different lines that intersect at point E, point F, or point G. Name the four angles formed by these two lines. Describe any relationships you see between any pair of angles formed by these two lines.

3. Compare your answers to Questions 1 and 2. Make a conjecture about the angles formed by intersecting lines.

LESSON 1.6

Technology Activity Keystrokes
For use with page 43

Keystrokes for Technology Activity

TI-92

Construct

1. Construct line *AB*

    [F2] 4 [ENTER] A (Move cursor to new location) [ENTER]

    [F2] 2 (Locate position for point *B*) [ENTER] B

2. Construct line *CD*

    [F2] 4 (Move cursor to location for point *C*) [ENTER] C

    (Move cursor to desired location for line) [ENTER]

    [F2] 2 (Locate position for point *D*) [ENTER] D

3. Construct intersection point *E*

    [F2] 3 (Move cursor to point at the intersection) [ENTER] E

Investigate

1. Measure the four angles formed by the intersecting lines (angles *AEC*, *AED*, *BEC*, and *BED*)

    [F6] 3 (Move cursor to point *A*) [ENTER] (Move cursor to point *E*)

    [ENTER] (Move cursor to point *C*) [ENTER]

    Repeat this procedure for the remaining angles.

2. Move the lines into a different position

    [F1] 1 (Select line *AB* or line *CD*) [ENTER]

Use the drag key 👆 and the cursor pad to drag the vertex and change the triangle.

Investigate

4. Calculate the sum of any two adjacent angles

    [F6] 6 (Use cursor to highlight an angle) [ENTER] [+]

    (Use cursor to highlight an adjacent angle) [ENTER] [ENTER]

    (the sum will appear on the screen)

5. Move the lines into a different position

    [F1] 1 (Select line *AB* or line *CD*) [ENTER] (Select Drag Key and move cursor to desired location)

    Calculate the sum of the two adjacent angles previously selected

    [F6] 6 (Use cursor to highlight an angle) [ENTER] [+]

    (Use cursor to highlight and adjacent angle) [ENTER] [ENTER]

    (the sum will appear on the screen)

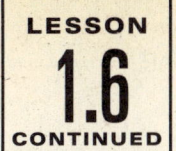

LESSON 1.6 CONTINUED

NAME _____ DATE _____

Technology Activity Keystrokes

For use with page 43

SKETCHPAD

Construct

1. Construct line *AB*. Choose the straightedge tool and make sure the straightedge tool symbol is a line. Draw line *AB*.

2. Construct line *CD*. Draw line *CD* so that it intersects *AB*.

3. Construct intersection point *E*. Choose the point tool, move the cursor until "point at intersection" is displayed in the dialog box in the left hand corner of the screen, and then click the mouse button.

Investigate

1. Measure the four angles formed by the intersecting lines (angles *AEC*, *AED*, *BEC*, and *BED*). To measure angle *AEC*, choose the pointer tool, select point *A*. Then hold down the shift key, select points *E* and *C*, and select **Angle** from the **Measure** menu. Repeat this procedure for the remaining angles. Before selecting the next angle, be sure to click anywhere in the work area to deselect the previous points.

2. Move the lines into a different position. With the pointer tool selected, drag point *A*.

Investigate

4. Calculate the sum of any two adjacent angles. Select **Calculate** from the **Measure** menu. Select the measure of one angle. Click ⊞ . Select the measure of the other angle. Click OK.

5. Move the lines into a different position. With the pointer tool selected, drag point *A*.

LESSON 1.6

Practice A
For use with pages 44–50

Use the figure at the right.

1. Are ∠1 and ∠2 adjacent?
2. Are ∠1 and ∠2 a linear pair?
3. Are ∠3 and ∠4 a linear pair?
4. Are ∠2 and ∠5 vertical angles?
5. Are ∠1 and ∠4 vertical angles?
6. Are ∠3 and ∠5 vertical angles?

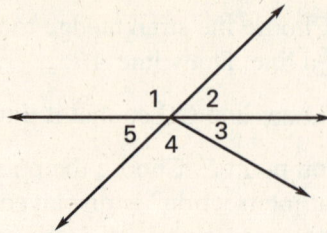

Use the figure at the right.

7. If $m\angle 6 = 78°$, then $m\angle 7 = $ ___?___ .
8. If $m\angle 8 = 94°$, then $m\angle 6 = $ ___?___ .
9. If $m\angle 9 = 124°$, then $m\angle 8 = $ ___?___ .
10. If $m\angle 7 = 47°$, then $m\angle 9 = $ ___?___ .
11. If $m\angle 8 = 158°$, then $m\angle 9 = $ ___?___ .
12. If $m\angle 7 = 15°$, then $m\angle 6 = $ ___?___ .

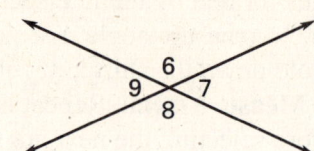

In Exercises 13–16, assume ∠A and ∠B are complementary and ∠B and ∠C are supplementary.

13. If $m\angle A = 42°$, then $m\angle B = $ ___?___ and $m\angle C = $ ___?___ .
14. If $m\angle B = 78°$, then $m\angle A = $ ___?___ and $m\angle C = $ ___?___ .
15. If $m\angle A = 17°$, then $m\angle B = $ ___?___ and $m\angle C = $ ___?___ .
16. If $m\angle B = 45°$, then $m\angle A = $ ___?___ and $m\angle C = $ ___?___ .

Find the value of the variable.

17.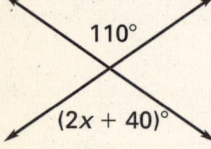

18.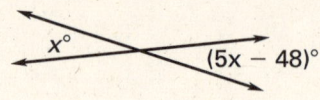

19.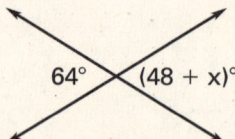

20.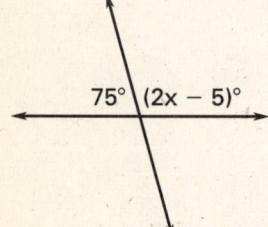

21.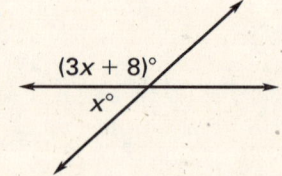

22.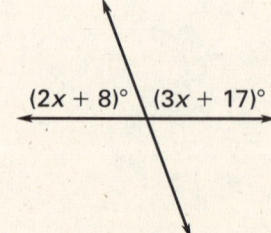

Lesson 1.6 Practice B

For use with pages 44–50

Use the figure at the right.

1. Are ∠1 and ∠2 a linear pair?
2. Are ∠4 and ∠5 a linear pair?
3. Are ∠3 and ∠1 vertical angles?
4. Are ∠2 and ∠5 vertical angles?

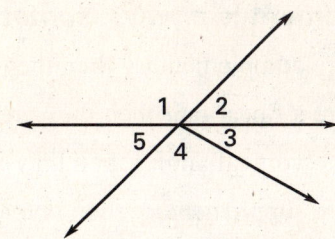

Use the figure at the right.

5. If $m\angle 6 = 51°$, then $m\angle 7 = $? .
6. If $m\angle 8 = 103°$, then $m\angle 6 = $? .
7. If $m\angle 9 = 136°$, then $m\angle 8 = $? .
8. If $m\angle 7 = 53°$, then $m\angle 9 = $? .

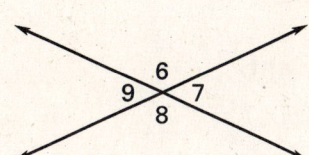

In Exercises 9–12, assume ∠A and ∠B are complementary and ∠B and ∠C are supplementary.

9. If $m\angle A = 48°$, then $m\angle B = $? and $m\angle C = $? .
10. If $m\angle B = 83°$, then $m\angle A = $? and $m\angle C = $? .
11. If $m\angle C = 127°$, then $m\angle B = $? and $m\angle A = $? .
12. If $m\angle A = 25°$, then $m\angle B = $? and $m\angle C = $? .

Find the value(s) of the variable(s).

13.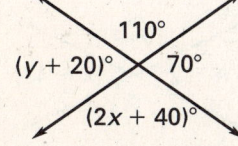
14.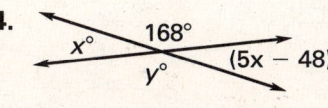
15.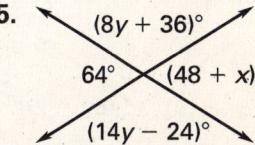

16.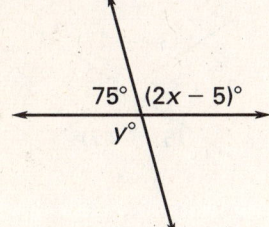
17.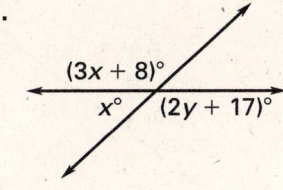
18.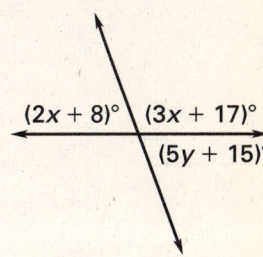

In Exercises 19 and 20, assume that ∠A is supplementary to ∠B and complementary to ∠C. Determine $m\angle A$, $m\angle B$, and $m\angle C$.

19. $m\angle A = x°$, $m\angle B = (x + 40)°$, $m\angle C = (x - 50)°$
20. $m\angle A = x°$, $m\angle B = (2x)°$, $m\angle C = (x - 30)°$

LESSON 1.6

Practice C
For use with pages 44–50

Decide if the statement is *always*, *sometimes*, or *never* true.

1. If two angles are complementary then they are adjacent.
2. If two angles are a linear pair then they are adjacent.
3. If two angles are vertical angles then they are adjacent.
4. If two angles are supplementary then one angle is acute and one angle is obtuse.

Use the figure at the right. Given one angle measure, find the other three.

5. If $m\angle 6 = 38°$
6. If $m\angle 8 = 84°$
7. If $m\angle 9 = 136°$
8. If $m\angle 7 = 27°$

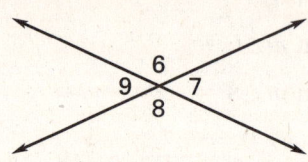

In Exercises 9–12, assume $\angle A$ and $\angle B$ are complementary and $\angle B$ and $\angle C$ are supplementary.

9. If $m\angle A = 52°$, then $m\angle B = \underline{\ ?\ }$ and $m\angle C = \underline{\ ?\ }$.
10. If $m\angle B = 67°$, then $m\angle A = \underline{\ ?\ }$ and $m\angle C = \underline{\ ?\ }$.
11. If $m\angle C = 107°$, then $m\angle B = \underline{\ ?\ }$ and $m\angle A = \underline{\ ?\ }$.
12. If $m\angle B = 12°$, then $m\angle A = \underline{\ ?\ }$ and $m\angle C = \underline{\ ?\ }$.

Find the value(s) of the variable(s).

13.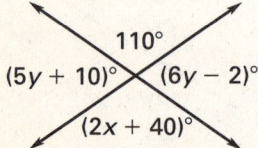

14.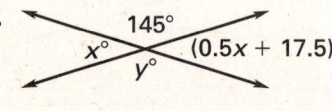

15.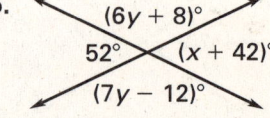

16.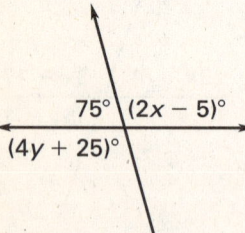

17.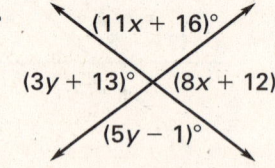

18.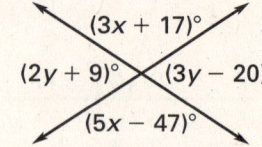

In Exercises 19 and 20, assume that $\angle A$ is supplementary to $\angle B$ and complementary to $\angle C$. Determine $m\angle A$, $m\angle B$, and $m\angle C$.

19. $m\angle A = (x + 10)°$, $m\angle B = (12x + 1)°$, $m\angle C = (5x + 2)°$
20. $m\angle A = (2.5x + 17)°$, $m\angle B = (21x - 25)°$, $m\angle C = (8x - 11)°$

LESSON 1.6

NAME _____ DATE _____

Reteaching with Practice

For use with pages 44–50

GOAL Identify vertical angles and linear pairs and identify complementary and supplementary angles

VOCABULARY

Two angles are **vertical angles** if their sides form two pairs of opposite rays.

Two adjacent angles are a **linear pair** if their noncommon sides are opposite rays.

Two angles are **complementary angles** if the sum of their measures is 90°. Each angle is the **complement** of the other.

Two angles are **supplementary angles** if the sum of their measures is 180°. Each angle is the **supplement** of the other.

EXAMPLE 1 *Identifying Vertical Angles and Linear Pairs*

a. Are ∠1 and ∠3 vertical angles?

b. Are ∠2 and ∠4 a linear pair?

c. Are ∠1 and ∠4 a linear pair?

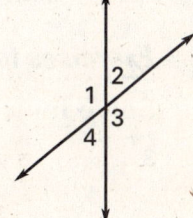

SOLUTION

a. Yes. The sides of the angles form two pairs of opposite rays.

b. No. The angles are not adjacent.

c. Yes. The angles are adjacent and their noncommon sides are opposite rays.

Exercises for Example 1

Use the figure to answer the questions.

1.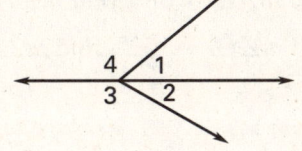

   a. Are ∠1 and ∠2 a linear pair?
   b. Are ∠1 and ∠3 vertical angles?
   c. Are ∠1 and ∠4 a linear pair?
   d. Are ∠2 and ∠4 vertical angles?

2.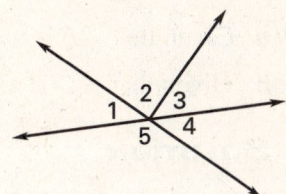

   a. Are ∠1 and ∠5 a linear pair?
   b. Are ∠1 and ∠2 a linear pair?
   c. Are ∠1 and ∠4 vertical angles?
   d. Are ∠3 and ∠5 vertical angles?

LESSON 1.6 CONTINUED

NAME _____ DATE _____

Reteaching with Practice

For use with pages 44–50

EXAMPLE 2 Finding Angle Measures

Solve for x in the diagram at the right.
Then find the angle measures.

SOLUTION

Use the fact that vertical angles are congruent.

$(7x - 25)° = (5x + 15)°$

$x = 20$

Use substitution to find the angle measures.

$m\angle FHI = (7x - 25)° = (7 \cdot 20 - 25)° = 115°$

$m\angle GHJ = (5x + 15)° = (5 \cdot 20 + 15)° = 115°$

Next, realize that $\angle FHI$ and $\angle FHG$ are a linear pair. So, the measures of these two angles must sum to 180°. So, $m\angle FHG = 180° - 115°$, so $m\angle FHG = 65°$.

Finally, notice that $\angle FHG$ and $\angle IHJ$ are vertical angles. So, $m\angle IHJ = 65°$.

Exercises for Example 2

Solve for x and y, then find the angle measures.

3.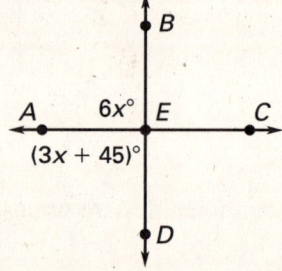

4.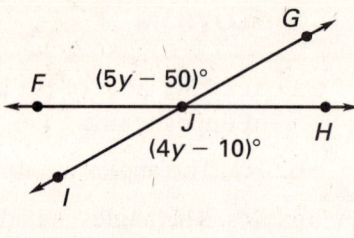

EXAMPLE 3 Finding Measures of Complements and Supplements

a. Given that $\angle E$ is a complement of $\angle F$ and $m\angle E = 68°$, find $m\angle F$.

b. Given that $\angle G$ is a supplement of $\angle H$ and $m\angle G = 152°$, find $m\angle H$.

SOLUTION

a. $m\angle F = 90° - m\angle E = 90° - 68° = 22°$

b. $m\angle H = 180° - m\angle G = 180° - 152° = 28°$

Exercises for Example 3

Find the measure of the angle.

5. Given that $\angle A$ is a complement of $\angle B$ and $m\angle B = 81°$, find $m\angle A$.

6. Given that $\angle C$ is a supplement of $\angle D$ and $m\angle C = 27°$, find $m\angle D$.

LESSON 1.6

NAME _____ DATE _____

Quick Catch-Up for Absent Students
For use with pages 43–50

The items checked below were covered in class on (date missed) _____

Activity 1.6: Angles and Intersecting Lines (p. 43)

____ **Goal:** Use goemetry software to construct intersecting lines and measure angles formed by the lines.

____ Student Help: Software Help

Lesson 1.6: Angle Pair Relationships

____ **Goal 1:** Identify vertical angles and linear pairs. (pp. 44–45)

Material Covered:

    ____ Example 1: Identifying Vertical Angles and Linear Pairs

    ____ Example 2: Finding Angle Measures

    ____ Example 3: Finding Angle Measures

Vocabulary:

    vertical angles, p. 44   linear pair, p. 44

____ **Goal 2:** Identifying complementary and supplementary angles. (p. 46)

Material Covered:

    ____ Student Help: Study Tip

    ____ Example 4: Identifying Angles

    ____ Example 5: Finding Measures of Complements and Supplements

    ____ Example 6: Finding the Measure of a Complement

Vocabulary:

    complementary angles, p. 46   complement, p. 46

    supplementary angles, p. 46   supplement, p. 46

____ Other (specify) _____

_____

Homework and Additional Learning Support

    ____ Textbook (specify) <u>pp. 47–50</u> _____

    ____ Internet: Extra Examples at www.mcdougallittell.com

    ____ *Reteaching with Practice* worksheet (specify exercises) _____

    ____ *Personal Student Tutor* for Lesson 1.6

LESSON 1.6

Interdisciplinary Application

For use with pages 44–50

Refraction

PHYSICS As you get ready to leave the swimming pool, a friend bumps you and your keys drop to the bottom of the pool's shallow end. You kneel down and stick your arm in to retrieve the keys, but find that they are not in the spot where they appeared to be. You have encountered the phenomenon of refraction. Refraction is the bending of a ray of light as it passes from one material to another, such as air and water. As the density of a material increases, it slows down the speed of light and changes the angle at which the light ray passes through the material.

The light ray from your eye passed through the air to the surface of the pool and then through the water to the bottom of the pool. To your eye, it appeared that the keys should be located at point E on the bottom of the pool in the diagram below. So, if they are not there, where are they?

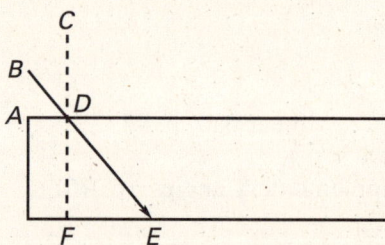

You kneel on the side of the pool at point A. Point B is your shoulder. $\vec{BE}$ represents your arm reaching into the pool. If your arm met the water at a 40° angle, then $m\angle ADB = 40°$.

1. Find the measure of $\angle BDC$. This is called the angle of incidence.
2. Find the measure of $\angle FDE$.
3. You can use Snell's Law from Physics to compute the angle of refraction for water. Let K be the point on the bottom of the pool where the keys really are. The angle of refraction would be $\angle FDK$.

    Snell's Law: $n_i \sin \theta_i = n_r \sin \theta_r$

    where n_i is the index of refraction for the incident material (air), n_r is the index of refraction for the refraction material (water), and θ_i and θ_r represent the angles of incident and refraction, respectively.

    The index of refraction for water is 1.333 and the index of refraction for air is 1.0003.

    Use the simplified equation

    $$\theta_r = \sin^{-1}\left(\frac{n_i \sin \theta_i}{n_r}\right)$$

    and the graphing calculator keystrokes [2nd] [SIN] $\left(\frac{n_i \sin \theta_i}{n_r}\right)$ to find the angle of refraction. (*Note:* Your graphing calculator needs to be in degree mode.)

4. Use Exercise 3 to determine if the keys are closer to you than you thought or farther away.

Lesson 1.6

Challenge: Skills and Applications
For use with pages 44–50

1. Explain what is wrong with the following argument: Note that ∠1 and ∠3 are vertical angles, and ∠2 and ∠4 are also vertical angles. Since ∠1 is a vertical angle and ∠2 is a vertical angle, and vertical angles are congruent, we may conclude that ∠1 ≅ ∠2.

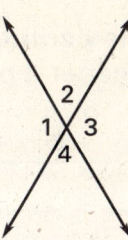

In Exercises 2–5, tell whether the statement is *true* or *false*. If it is true, explain why; if it is false, sketch a counterexample.

2. Given ∠ACB and ∠BCD with a common side $\overrightarrow{CB}$:
   If ∠ACD is a right angle, then ∠ACB and ∠BCD are complementary.

3. Given ∠ACB and ∠BCD with a common side $\overrightarrow{CB}$:
   If ∠ACD is a straight angle, then ∠ACB and ∠BCD are supplementary.

4. If ∠RUS and ∠SUT are a linear pair, and ∠SUT and ∠TUV are also a linear pair, then ∠RUS and ∠TUV are vertical angles.

5. If ∠RUS and ∠TUV are vertical angles, then ∠RUS and ∠SUT are a linear pair.

In Exercises 6–11, find the values of *x* and *y* in the diagram.

6.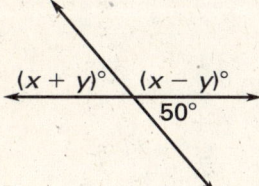

7.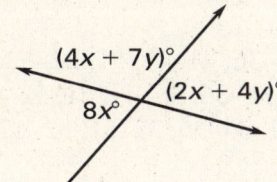

8.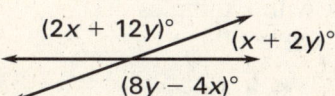

9. (6x + y)°, (9y − 2x)°, (2x − y)°, (8y − 2x)°

10.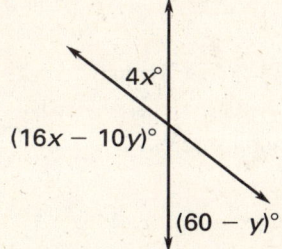

11.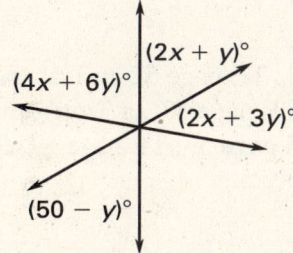

LESSON 1.7

Teacher's Name _____ Class _____ Room _____ Date _____

Lesson Plan
2-day lesson (See *Pacing the Chapter*, TE pages 1C–1D)   For use with pages 51–58

GOALS
1. Find the perimeter and area of common plane figures.
2. Use a general problem-solving plan.

State/Local Objectives _____
_____

✓ **Check the items you wish to use for this lesson.**

STARTING OPTIONS
___ Homework Check: TE page 47: Answer Transparencies
___ Warm-Up or Daily Homework Quiz: TE pages 51 and 50, CRB page 94, or Transparencies

TEACHING OPTIONS
___ Motivating the Lesson: TE page 52
___ Lesson Opener (Activity): CRB page 95 or Transparencies
___ Technology Activity with Keystrokes: CRB pages 96–99
___ Examples: Day 1: 1–3: SE pages 51–52; Day 2: 4–6, SE pages 53–54
___ Extra Examples: Day 1: TE page 52 or Transp.; Day 2: TE pages 53–54 or Transp.
___ Closure Question: TE page 54
___ Guided Practice: SE page 55   Day 1: Exs. 1–5; Day 2: Exs. 6–8

APPLY/HOMEWORK
Homework Assignment
___ Basic   Day 1: 9–33; Day 2: 34–36, 41–49, 51, 52–62 even; Quiz 3: 1–10
___ Average   Day 1: 9–33; Day 2: 34–36, 38–49, 51, 52–62 even; Quiz 3: 1–10
___ Advanced   Day 1: 9–33; Day 2: 34–36, 38–51, 52–62 even; Quiz 3: 1–10

Reteaching the Lesson
___ Practice Masters: CRB pages 100–102 (Level A, Level B, Level C)
___ Reteaching with Practice: CRB pages 103–104 or Practice Workbook with Examples
___ Personal Student Tutor

Extending the Lesson
___ Cooperative Learning Activity: CRB page 106
___ Applications (Real-Life): CRB page 107
___ Challenge: SE page 57; CRB page 108 or Internet

ASSESSMENT OPTIONS
___ Checkpoint Exercises: Day 1: TE page 52 or Transp.; Day 2: TE pages 53–54 or Transp.
___ Daily Homework Quiz (1.7): TE page 57, or Transparencies
___ Standardized Test Practice: SE page 57; TE page 57; STP Workbook; Transparencies
___ Quiz (1.6–1.7): SE page 58

Notes _____
_____
_____

92 **Geometry**
Chapter 1 Resource Book

Copyright © McDougal Littell Inc.
All rights reserved.

LESSON 1.7

TEACHER'S NAME _____ CLASS _____ ROOM _____ DATE _____

Lesson Plan for Block Scheduling
1-day lesson (See *Pacing the Chapter,* TE pages 1C–1D)   For use with pages 51–58

GOALS
1. Find the perimeter and area of common plane figures.
2. Use a general problem-solving plan.

State/Local Objectives _____
_____
_____

CHAPTER PACING GUIDE	
Day	Lesson
1	1.1 (all); 1.2 (all)
2	1.3 (all)
3	1.4 (all); 1.5 (begin)
4	1.5 (end); 1.6 (begin)
5	1.6 (end); **1.7 (begin)**
6	**1.7 (end)**; Review Ch. 1
7	Assess Ch. 1; 2.1 (all)

✓ Check the items you wish to use for this lesson.

STARTING OPTIONS
____ Homework Check: TE page 47: Answer Transparencies
____ Warm-Up or Daily Homework Quiz: TE pages 51 and 50, CRB page 94, or Transparencies

TEACHING OPTIONS
____ Motivating the Lesson: TE page 52
____ Lesson Opener (Activity): CRB page 95 or Transparencies
____ Technology Activity with Keystrokes: CRB pages 96–99
____ Examples: Day 5: 1–3: SE pages 51–52; Day 6: 4–6, SE pages 53–54
____ Extra Examples: Day 5: TE page 52 or Transp.; Day 6: TE pages 53–54 or Transp.
____ Closure Question: TE page 54
____ Guided Practice: SE page 55   Day 5: Exs. 1–5; Day 6: Exs. 6–8

APPLY/HOMEWORK
Homework Assignment (See also the assignment for Lesson 1.6.)
____ Block Schedule: Day 5: 9–33; Day 6: 34–36, 38–49, 51, 52–62 even; Quiz 3: 1–10

Reteaching the Lesson
____ Practice Masters: CRB pages 100–102 (Level A, Level B, Level C)
____ Reteaching with Practice: CRB pages 103–104 or Practice Workbook with Examples
____ Personal Student Tutor

Extending the Lesson
____ Cooperative Learning Activity: CRB page 106
____ Applications (Real-Life): CRB page 107
____ Challenge: SE page 57; CRB page 108 or Internet

ASSESSMENT OPTIONS
____ Checkpoint Exercises: Day 5: TE page 52 or Transp.; Day 6: TE pages 53–54 or Transp.
____ Daily Homework Quiz (1.7): TE page 57, or Transparencies
____ Standardized Test Practice: SE page 57; TE page 57; STP Workbook; Transparencies
____ Quiz (1.6–1.7): SE page 58

Notes _____
_____
_____

LESSON 1.7

WARM-UP EXERCISES
For use before Lesson 1.7, pages 51–58

Find the distance between each pair of points in the coordinate plane.

1. $(3, 1)$ and $(3, -5)$

2. $(-2, 0)$ and $(4, 2)$

3. $(-1, -5)$ and $(3, -5)$

4. $(6, -4)$ and $(4, -2)$

DAILY HOMEWORK QUIZ
For use after Lesson 1.6, pages 43–50

Use the figure to answer the questions.

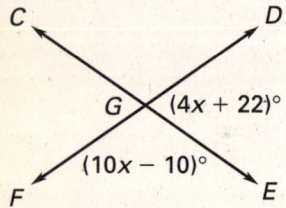

1. Are angles $\angle CGF$ and $\angle FGE$ vertical angles?

2. Are angles $\angle CGF$ and $\angle FGE$ a linear pair?

3. If $m\angle CGD = 120°$, find $m\angle CGF$.

4. Find the value of x.

$\angle A$ and $\angle B$ are complementary. Find $m\angle A$ and $m\angle B$.

5. $m\angle A = 7x + 1$, $m\angle B = 4x + 1$

LESSON 1.7

Activity Lesson Opener

For use with pages 51–58

SET UP: Work with a partner.

YOU WILL NEED: • piece of string

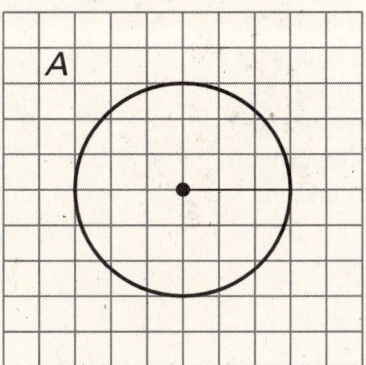
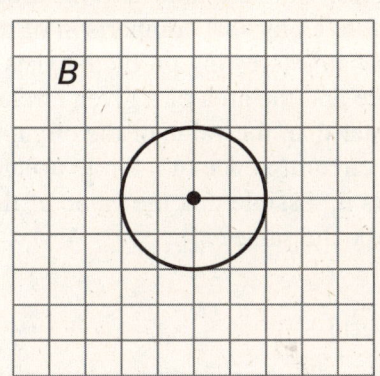
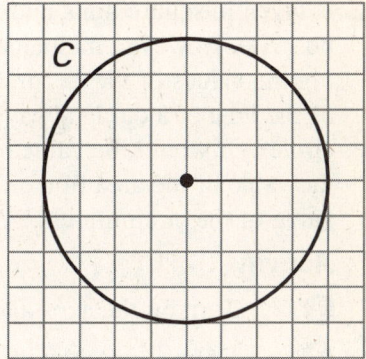

1. Find the diameter (d) and the radius (r) of each circle in grid units. Then calculate the radius squared (r^2) of each circle. Record your answers in the table below.

2. Use your piece of string to measure the circumference (C) of each circle. Give this length in grid units.

3. Estimate the area (A) of each circle by counting the grid squares inside the circle. Record your answer to the nearest square unit in the table below.

Circle	d	r	r^2	C	A
A					
B					
C					

4. Describe any relationships you see among the numbers in the columns in the table above.

LESSON 1.7

NAME _____ DATE _____

Technology Activity
For use with page 51–58

GOAL To determine the effects that the length of the radius of a circle has on the circumference and area of the circle

The circumference of a circle is given in linear units and the area of a circle is given in square units and therefore a physical comparison of the two is not possible. However, mathematically an interesting problem exists when comparing the values of the circumference and the area of a given circle. Is the value of the area of a circle always greater than the value of the circumference of the circle? Or would the value of the circumference of a circle usually be larger than the value of the area of a circle? Is it possible that the value of the area and the value of the circumference could be the same?

Activity

1. Turn on the axes and the grid.

2. Define a circle with center at the origin and a radius of one.

3. Measure and tabulate the radius, circumference, and the area of the circle.

4. Change the length of the radius to 2.0. Tabulate the radius, circumference, and the area of the circle.

5. Change the length of the radius to 3.0. Tabulate the radius, circumference, and the area of the circle.

6. Change the length of the radius to 4.0. Tabulate the radius, circumference, and the area of the circle.

Exercises

1. At what length did the radius produce equal values for both the circumference and the area of the circle?

2. Given that the formula for the circumference of a circle is $C = 2\pi r$ and that the area of a circle is $A = \pi r^2$, show algebraically that your answer for Exercise 1 is true. (*Hint*: Set the two formulas equal to each other and solve for r)

3. Make three statements relating the length of the radius of a circle to the value of the circumference and area of that circle.

LESSON 1.7 CONTINUED

NAME _____ DATE _____

Technology Activity Keystrokes

For use with page 51–58

TI-92

1. Turn on axes and the grid

   F8 9 (Set Coordinate Axes to RECTANGULAR and Grid to ON)
   ENTER

2. First define the radius with a segment of length one

   F2 5 (Move the cursor to a point on the grid) **ENTER** (Move the cursor to a second point on the grid that is two points away) **ENTER**

   Measure the length of the segment

   F6 1 (Place cursor on segment) **ENTER**

   Define a circle with the center at the origin and a radius of one

   F4 8 (Place cursor at the origin) **ENTER** (Select segment) **ENTER**

3. Measure the circumference, and the area of the circle

   F6 1 (Place cursor on the circle) **ENTER**

   F6 2 (Place cursor on the circle) **ENTER**

   Tabulate the radius, circumference, and the area of the circle

   F6 7 (Place cursor on the value of the radius) **ENTER** (Place cursor on the value of the circumference) **ENTER** (Place cursor on the value of the area) **ENTER**

   F6 7 1 (The data will be stored into the variable sysData and can be seen at any time by selecting **APPS** 6 2 Type should be Data, Folder should be main and Variable should be sysData. Place cursor on sysData and press **ENTER**.)

4. Change the length of the radius to 2.0. Then tabulate the radius, circumference, and the area of the circle.

   F1 1 (Place cursor on endpoint of segment and use the drag key and the cursor pad to drag the endpoint of the radius)

   F6 7 **2** (The three values should be highlighted)

   F6 7 **1** (The three values should now be tabulated)

5. Repeat step 4 for a radius of 3.0.

6. Repeat step 4 for a radius of 4.0.

LESSON 1.7 CONTINUED

Technology Activity Keystrokes

For use with page 51–58

SKETCHPAD

1. Turn on the axes and the grid. Select **Snap to Grid** from the **Graph** menu.

2. Define a circle with the center at the origin and a radius of one. Choose the compass tool from the toolbox. Click and hold down the mouse button on point *A*. Extend the circle until the radius is one unit.

3. Measure the radius, circumference, and the area of the circle. Choose the translate selection arrow tool. Then select the circle and choose **Radius** from the **Measure** menu. Choose **Circumference** from the **Measure** menu. Choose **Area** from the **Measure** menu. Tabulate the radius, circumference, and the area of the circle. Use the mouse to select the radius, circumference, and area measures. Choose **Tabulate** from the **Measure** menu.

4. Change the length of the radius by dragging point *C* until the radius is 2.0. Tabulate the radius, circumference, and area of the circle by double clicking the table.

5. Repeat Step 4 for a radius of 3.0.

6. Repeat Step 4 for a radius of 4.0.

LESSON 1.7

NAME _____ DATE _____

Technology Activity Keystrokes
For use with page 56

Keystrokes for Exercise 37

EXCEL

Select cell A1.

Length `ENTER` Width `ENTER` Area `ENTER` Perimeter `ENTER`

Select cell B1

1 `TAB` = B1 + 1 `ENTER`

Select cell C1. From the **Edit** menu, choose **Copy**.

Select cells D1–CW1. From the **Edit** menu, choose **Paste**.

Select cell B2.

= 100/B1 `ENTER`

Select cell B2. From the **Edit** menu, choose **Copy**.

Select cells C2–CW2. From the **Edit** menu, choose **Paste**.

Select cell B3.

= B1*B2 `ENTER`

Select cell B3. From the **Edit** menu, choose **Copy**.

Select cells C3–CW3. From the **Edit** menu, choose **Paste**.

Select cell B4.

= 2*B1 + 2*B2 `ENTER`

Select cell B4. From the **Edit** menu, choose **Copy**.

Select cells C4–CW4. From the **Edit** menu, choose **Paste**.

LESSON 1.7

Practice A
For use with pages 51–58

Find the perimeter (or circumference) and area of the figure. (Where necessary, use $\pi \approx 3.14$.)

1.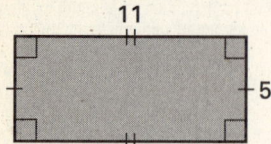
2.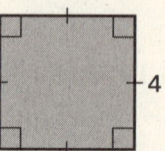
3.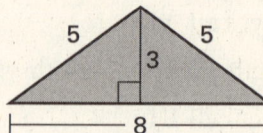
4.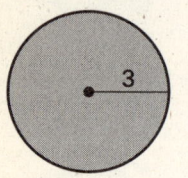
5.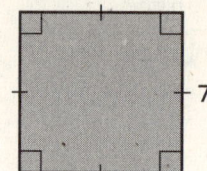
6.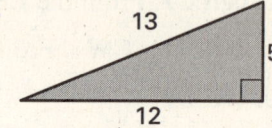
7.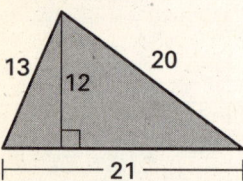
8.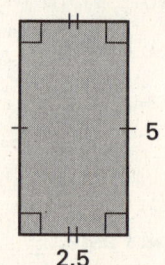
9.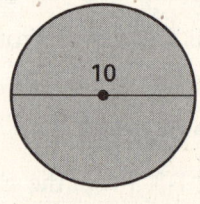

Find the area of the figure described.

10. Rectangle with length 8 centimeters and width 4.5 centimeters
11. Triangle with height 5 inches and base 16 inches
12. Circle with diameter 12 feet (use $\pi \approx 3.14$)

Find the area of the figure. (Where necessary, use $\pi \approx 3.14$.)

13.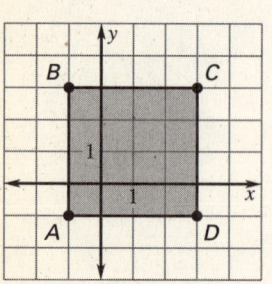
14.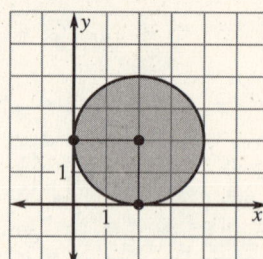
15.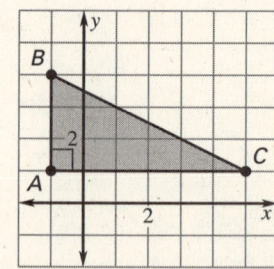

LESSON 1.7

Practice B
For use with pages 51–58

Find the perimeter (or circumference) and area of the figure. (Where necessary, use $\pi \approx 3.14$.)

1.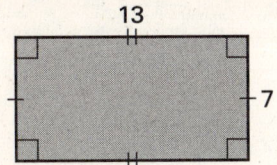
2.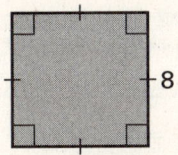
3.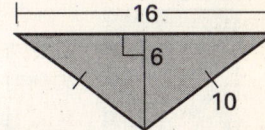
4.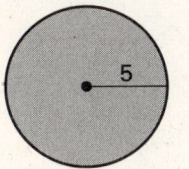
5.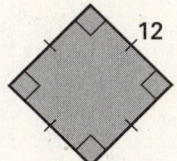
6.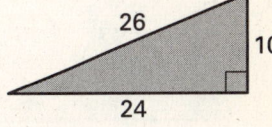
7.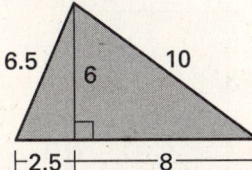
8.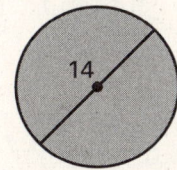
9.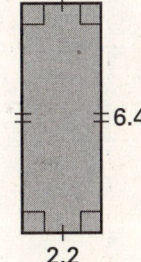

Find the area of the figure described.

10. Rectangle with length 12.3 centimeters and width 5 centimeters
11. Triangle with height 7 inches and base 14.4 inches
12. Circle with diameter 40 feet (use $\pi \approx 3.14$)

Find the area of the figure. (Where necessary, use $\pi \approx 3.14$.)

13.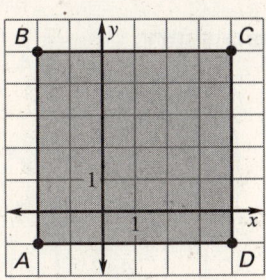
14.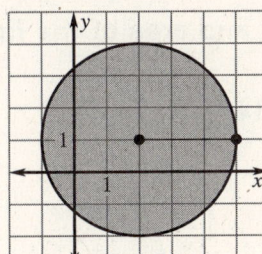
15.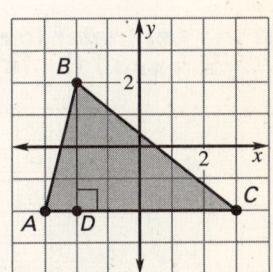

LESSON 1.7

Practice C
For use with pages 51–58

Find the perimeter (or circumference) and area of the figure. (Where necessary, use $\pi \approx 3.14$.)

1.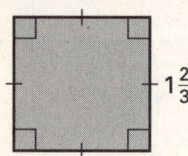

2.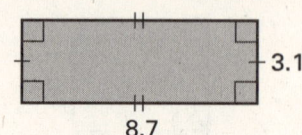

3.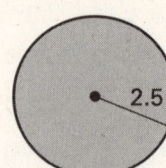

4.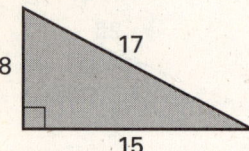

5.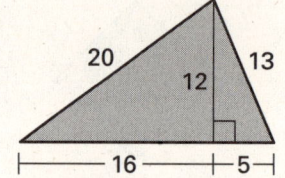

6.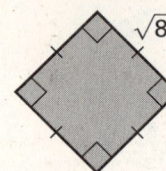

7.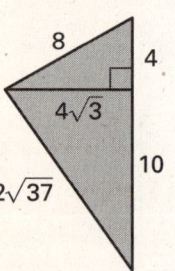

8.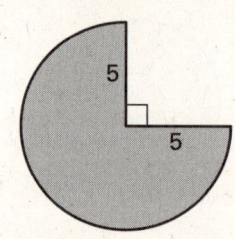

9.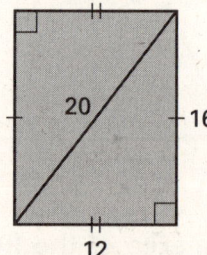

Find the area of the figure described.

10. Rectangle with length 6.2 centimeters and width 4.3 centimeters

11. Triangle with height 9 inches and base 15 inches

12. Circle with diameter 50 feet (use $\pi \approx 3.14$)

Find the perimeter (circumference) and area of the figure. (Where necessary, use $\pi \approx 3.14$.)

13.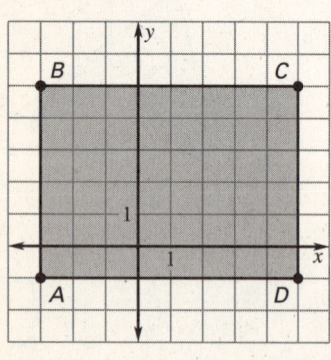

14.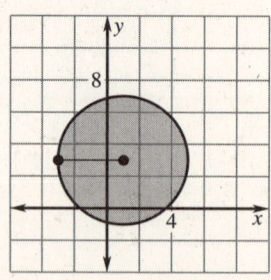

15.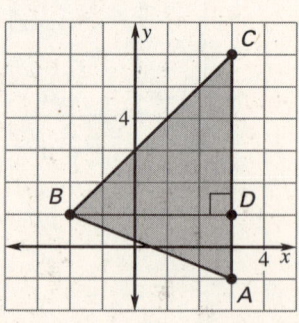

16. **Shingles** You are buying shingles for a roof. Each bundle of shingles will cover 27 square feet. The roof consists of two rectangular parts, and each is 70 feet by 30 feet. How many bundles of shingles do you need?

17. **Irrigation** A new irrigation system has been installed. Each irrigation arm covers a circular region with a radius of 35 feet. How many square feet will 4 irrigation arms cover?

LESSON 1.7

Reteaching with Practice

For use with pages 51–58

GOAL Find the perimeter and area of common plane figures and use a general problem-solving plan

VOCABULARY

Formulas for the perimeter P, area A, and circumference C of some common plane figures are given below.

Square

Side length s

$P = 4s$

$A = s^2$

Rectangle

length l and width w

$P = 2l + 2w$

$A = lw$

Triangle

Side lengths a, b, and c,

base b, and height h

$P = a + b + c$

$A = \frac{1}{2}bh$

Circle

radius r

$C = 2\pi r$

$A = \pi r^2$

A Problem-Solving Plan:

1. Ask yourself what you need to solve the problem. Write a **verbal model** or **draw a sketch** that will help you find what you need to know.

2. **Label known and unknown facts** on or near your sketch.

3. Use labels and facts to **choose related definitions, theorems, formulas,** or other results you may need.

4. **Reason logically** to link the facts, using a proof or other written argument.

5. Write a **conclusion** that answers the original problem. **Check** that your reasoning is correct.

EXAMPLE 1 Finding the Perimeter and Area of a Square

Find the perimeter and area of a square with a side of 4 inches.

SOLUTION

Begin by drawing a diagram and labeling one of the sides. Then, use the formulas for perimeter and area of a square.

$P = 4s$ $A = s^2$

$ = 4(4)$ $ = 4^2$

$ = 16$ $ = 16$

So, the perimeter is 16 inches and the area is 16 square inches.

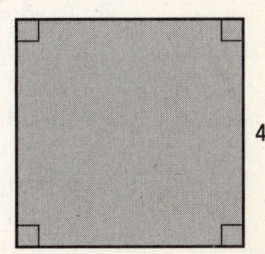

LESSON 1.7 CONTINUED

NAME _____ DATE _____

Reteaching with Practice

For use with pages 51–58

Exercises for Example 1

Find the perimeter (or circumference) and area of the figure. (Where necessary, use $\pi \approx 3.14$.)

1.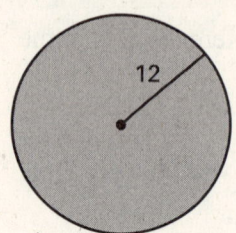

2.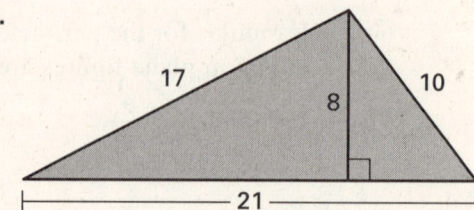

EXAMPLE 2 Using the Area of a Circle

You are making a cardboard model of a car. You make the tires with a radius of 18 centimeters. If the rim alone has a radius of 14 centimeters, what is the area of the rubber part of the tire?

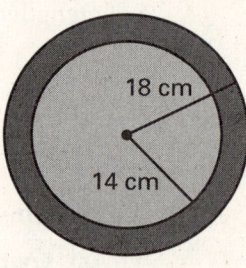

SOLUTION

Draw a Sketch From the diagram, you can see that the area of the rubber can be represented by the area of the larger circle minus the area of the smaller circle.

Verbal Model | Area of rubber | = | Area of large circle | − | Area of small circle |

Labels
Area of rubber = A (square centimeters)
Radius of whole tire = 18 (centimeters)
Radius of rim = 14 (centimeters)

Reasoning
$A = \pi \cdot 18^2 - \pi \cdot 14^2$ Write model for rubber area.
$\approx 3.14 \cdot 324 - 3.14 \cdot 196$ $\pi \approx 3.14$ and evaluate powers.
$= 1017.36 - 615.44$ Multiply.
$= 401.92$ Subtract.

The area of the rubber is about 401.92 square centimeters.

Exercise for Example 2

3. A window has the shape of a rectangle with a half-circle (see figure). The rectangle has a width of 3 feet and a height of 7 feet. Find the perimeter and area of the window. Use $\pi \approx 3.14$ where necessary.

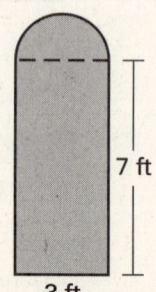

LESSON 1.7

NAME _____ DATE _____

Quick Catch-Up for Absent Students
For use with pages 51–58

The items checked below were covered in class on (date missed) _____

Lesson 1.7: Introduction to Perimeter, Circumference, and Area

____ **Goal 1:** Find the perimeter and area of common plane figures. (pp. 51–52)

Material Covered:

____ Example 1: Finding the Perimeter and Area of a Rectangle

____ Student Help: Study Tip

____ Example 2: Finding the Area and Circumference of a Circle

____ Student Help: Skills Review

____ Example 3: Finding Measurements of a Triangle in a Coordinate Plane

____ **Goal 2:** Use a general problem-solving plan. (pp. 53–54)

Material Covered:

____ Example 4: Using the Area of a Rectangle

____ Example 5: Using the Area of a Square

____ Example 6: Using the Area of a Triangle

____ Other (specify) _____
_____

Homework and Additional Learning Support

____ Textbook (specify) pp. 55–58 _____
_____

____ *Reteaching with Practice* worksheet (specify exercises) _____

____ *Personal Student Tutor* for Lesson 1.7

Lesson 1.7

NAME _____ DATE _____

Cooperative Learning Activity
For use with pages 51–58

GOAL To use formulas and a ruler to approximate areas and perimeters of irregularly shaped figures

Materials: ruler, pencil, calculator

Exploring Area and Perimeter

The area of a plane figure is the measure of the region enclosed by the figure. The perimeter of a polygon is the sum of the lengths of its sides. People in many occupations work with area and perimeter. For example, builders, painters, and interior decorators calculate area and perimeter to order materials.

Instructions

❶ Discuss the best strategy for each problem.

❷ Use your ruler to approximate the area of the following figure (in square centimeters).

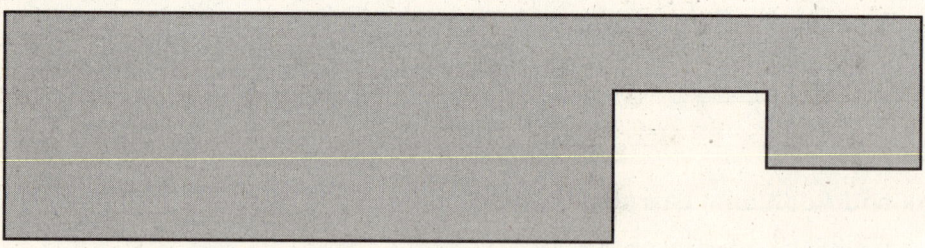

❸ Which of the figures below has the larger perimeter? What is the perimeter (in centimeters)? Which figure has the larger area? What is the area (in square centimeters)?

a.   b.

❹ Use your ruler to approximate the area of this dodecagon. Try doing this in the least number of measurements.

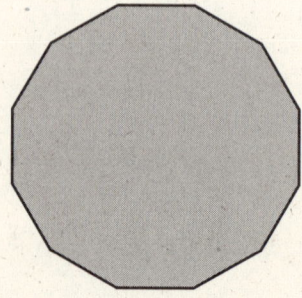

Analyzing the Results

1. What strategy did your group use to solve each problem?

2. Find the area and perimeter of your classroom. Compare your results with other groups.

106 Geometry
Chapter 1 Resource Book

LESSON 1.7

Real-Life Application: When Will I Ever Use This?

For use with pages 51–58

The Pentagon

The Pentagon in Washington, D.C. is the headquarters for the Department of Defense, and is one of the world's largest office buildings. It has three times the floor space of the Empire State Building in New York City. Although it was completed on January 15, 1943, it is still thought of as one of the most efficient office buildings in the world. It is a regular pentagon, which means that it has five congruent sides. The building consists of five concentric rings with ten spokelike corridors connecting each ring. It has 17.5 miles of corridors, but it only takes seven minutes to walk between any two points in the building. The length of each outer wall is 921 feet. It has 5 floors and a height of 77 feet 3.5 inches. You and your family travel to Washington, D.C. and decide to tour The Pentagon. Tours take approximately 90 minutes and require about a mile of walking.

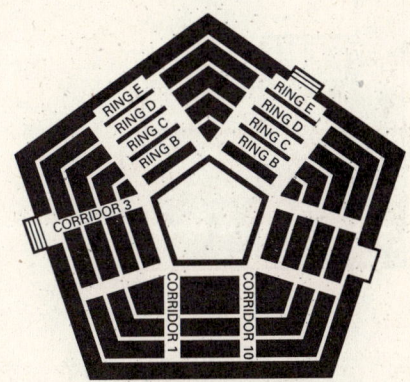

1. Will you get to walk around all five floors during your visit? (1 mile = 5280 feet)

2. Using the fact that distance equals rate times time, determine if your are walking fast or slow during the tour. The average walking speed for humans is about 3 feet per second.

3. The Pentagon building and inner courtyard (a smaller pentagon at the center of The Pentagon) cover 34 acres of land. The inner courtyard covers 217,800 square feet. How many square feet does The Pentagon cover? (1 acre = 43,560 square feet)

4. The Pentagon has 3,700,000 square feet of usable floor space, and the approximate cost of the building was $49.6 million. What was the cost per square foot?

LESSON 1.7

NAME _____ DATE _____

Challenge: Skills and Applications
For use with pages 51–58

In Exercises 1–8, find the perimeter (or circumference) and area of the figure. (Where necessary, use $\pi \approx 3.14$.)

1.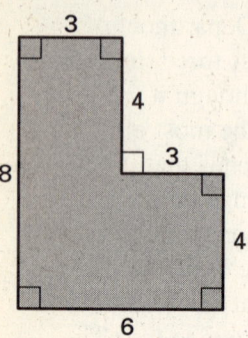

2.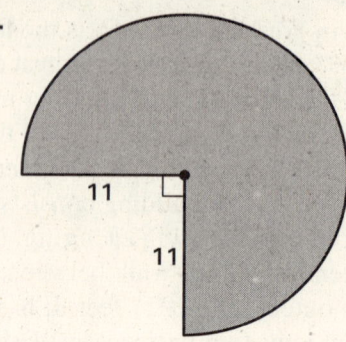

3.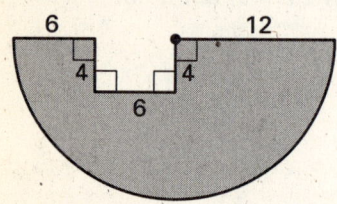

4.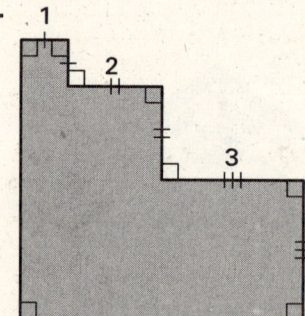

5.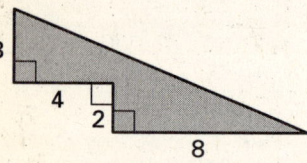

6.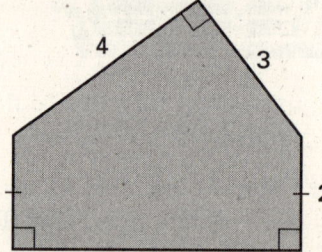

7.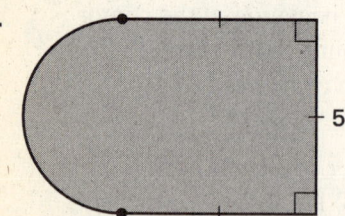

8.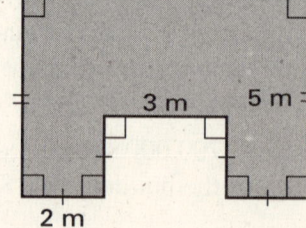

108 Geometry
Chapter 1 Resource Book

Chapter Review Games and Activities

For use after Chapter 1

What do you get if you divide the circumference of a jack o' lantern by its diameter?

The first letters of the answers to the following questions will spell out the answer to this riddle.

1. What is the name of an object with no dimension that is represented by a small dot?

2. What kind of a statement is a conjecture?

3. What is the point called that divides a line segment into two congruent segments?

4. What is the name of an object that extends in two dimensions and is represented by a shape that looks like a tabletop?

5. What is the name of a recreational toy that has a frame that bisects at least two of the angles it supports?

6. When two lines have a point in common, we say that they _____.

7. Three or more points that do not all lie on the same line are described as being _____.

8. What is the name of the theorem on which the distance formula is based?

9. The point A is the _____ point of the ray $\overrightarrow{AB}$.

CHAPTER 1

Name _____ Date _____

Chapter Test A

For use after Chapter 1

In Questions 1–3, predict the next three numbers.

1. 2, 5, 8, 11, ...
2. 1, 2, 4, 7, ...
3. 243, 81, 27, 9, ...

Use the diagram to name the figures.

4. Three collinear points
5. Three noncollinear points
6. Four noncoplanar points
7. Two intersecting lines

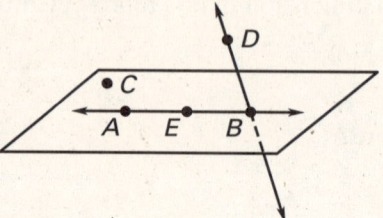

Find the length of the segment.

8. $\overline{AB}$
9. $\overline{CD}$

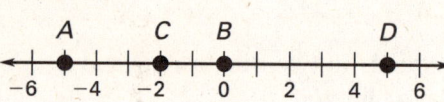

Use the diagram to find the measure of the angle.

10. $\angle AMC$
11. $\angle DMV$
12. $\angle AME$

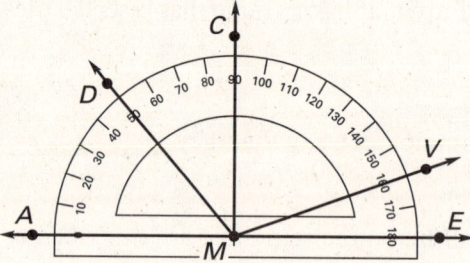

In Questions 13–15, refer to the diagram for Questions 10–12. State what type of angle is formed.

13. $\angle AMC$
14. $\angle DMV$
15. $\angle AME$

Find the coordinates of the midpoint of a segment with the given endpoints. Then find the length of the segment.

16. $A(0, 0)$,
    $B(0, 10)$
17. $C(2, 9)$,
    $D(-2, -9)$
18. $E(4, -8)$,
    $F(-6, 6)$

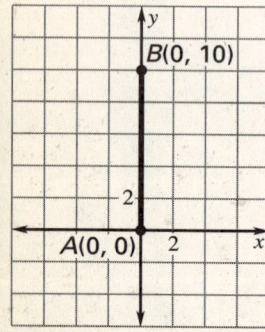
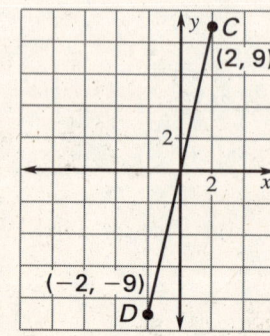
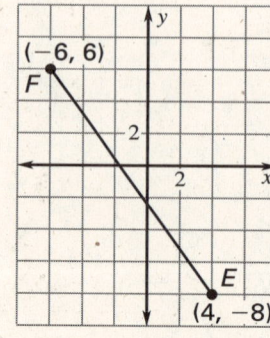

Answers

1. _____
2. _____
3. _____
4. _____
5. _____
6. _____
7. _____
8. _____
9. _____
10. _____
11. _____
12. _____
13. _____
14. _____
15. _____
16. midpoint: _____
    length: _____
17. midpoint: _____
    length: _____
18. midpoint: _____
    length: _____

Geometry
Chapter 1 Resource Book

Chapter Test A

For use after Chapter 1

In Questions 19 and 20, $\vec{EF}$ is the angle bisector of $\angle TEA$. Find the two angle measures not given in the diagram.

19.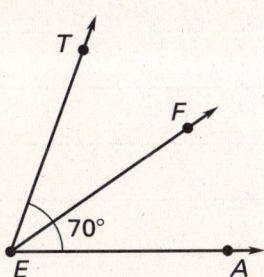
20.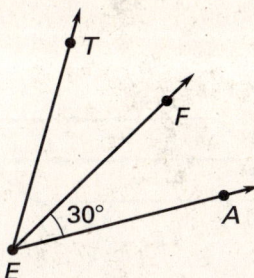

In Questions 21–22, $\vec{BD}$ bisects $\angle ABC$. Find the value of x.

21.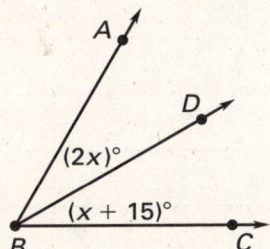
22.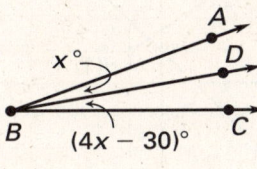

In Questions 23–25, use the diagram to solve for the missing angle measure.

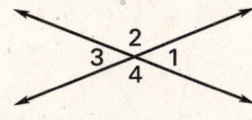

23. If $m\angle 1 = 25°$, then $m\angle 3 =$ ___?___.
24. If $m\angle 2 = 95°$, then $m\angle 1 =$ ___?___.
25. If $m\angle 4 = 130°$, then $m\angle 2 =$ ___?___.

Find the values of the variables.

26.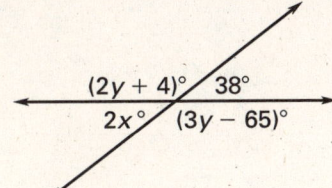
27.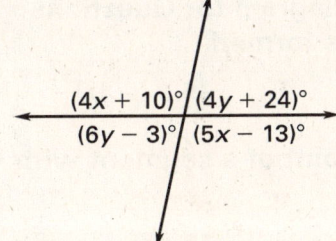

Find the area of the figures described.

28. Triangle with height 4 cm and base 5 cm
29. Square with side length 3 ft
30. Circle with radius 5 yd

19. _____
20. _____
21. _____
22. _____
23. _____
24. _____
25. _____
26. _____
27. _____
28. _____
29. _____
30. _____

CHAPTER 1

Chapter Test B
For use after Chapter 1

NAME _____ DATE _____

In Questions 1–3, predict the next three numbers.

1. $-2, 2, 6, 10, \ldots$ 2. $-5, -2, 4, 13, \ldots$ 3. $1, -\frac{1}{2}, \frac{1}{4}, -\frac{1}{8}, \ldots$

Use the diagram to name the figures.

4. Three collinear points
5. Three noncollinear points
6. Four noncoplanar points
7. Two intersecting lines

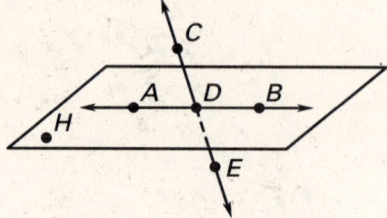

Find the length of the segment.

8. $\overline{AB}$
9. $\overline{CD}$

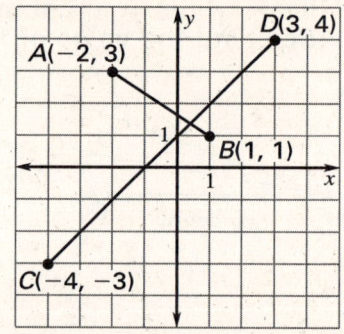

Use the diagram to find the measure of the angle.

10. $\angle EAB$
11. $\angle CAF$
12. $\angle BAD$

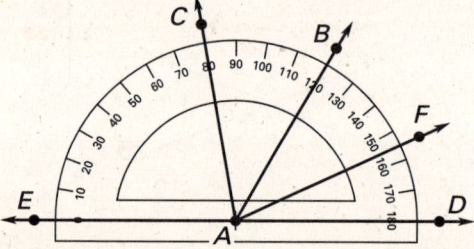

In Questions 13–15, refer to the diagram for Questions 10–12. State what type of angle is formed.

13. $\angle EAB$ 14. $\angle CAF$ 15. $\angle EAD$

Find the coordinates of the midpoint of a segment with the given endpoints.

16. $A(0, 0)$ 17. $C(2, 9)$ 18. $E(-3, -3)$
    $B(0, -12)$ $D(-2, -1)$ $F(9, -15)$

Answers

1. _____
2. _____
3. _____
4. _____
5. _____
6. _____
7. _____
8. _____
9. _____
10. _____
11. _____
12. _____
13. _____
14. _____
15. _____
16. _____
17. _____
18. _____

Chapter Test B
For use after Chapter 1

In Questions 19 and 20, $\overrightarrow{EF}$ is the angle bisector of $\angle TEA$. Find the two angle measures not given in the diagram.

19.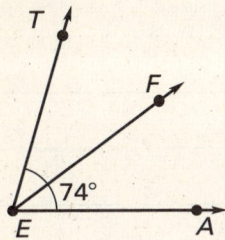

20.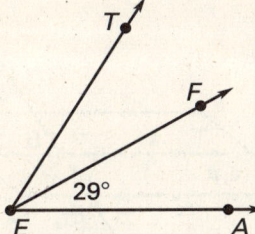

Answers
19. _____
20. _____
21. _____
22. _____
23. _____
24. _____
25. _____
26. _____
27. _____
28. _____
29. _____
30. _____
31. _____
32. _____
33. _____

In Questions 21–22, $\overrightarrow{BD}$ bisects $\angle ABC$. Find the value of x.

21.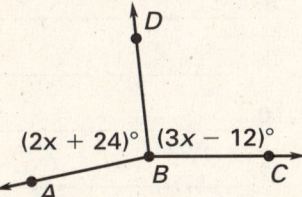

22.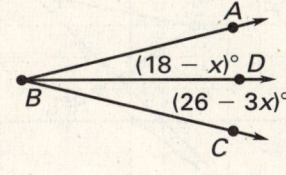

In Questions 23–25, use the diagram to solve for the missing angle measure.

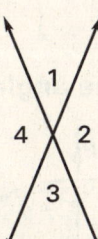

23. If $m\angle 1 = 44°$, then $m\angle 3 =$ ____.
24. If $m\angle 2 = 78°$, then $m\angle 1 =$ ____.
25. If $m\angle 4 = 33°$, then $m\angle 2 =$ ____.

In Questions 26–28, refer to the diagram for Questions 23–25 and state whether the given angles are a linear pair or are vertical angles.

26. $\angle 1$ and $\angle 3$ 27. $\angle 2$ and $\angle 1$ 28. $\angle 4$ and $\angle 2$

Find the values of the variables.

29.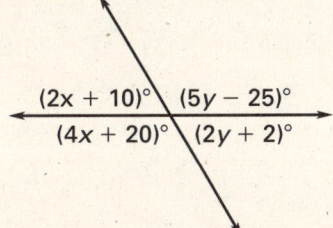

30.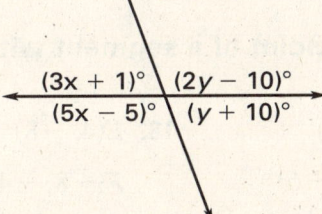

Find the area of the figure described.

31. Triangle with height 2.5 cm and base 8 cm
32. Square with side length 7 ft
33. Circle with radius 4 yd

Chapter 1

Chapter Test C
For use after Chapter 1

In Questions 1–3, predict the next three numbers.

1. 0.08, 0.4, 2, 10, ...
2. −2, −1, 1, 4, ...
3. −2, −7, −17, −32, ...

Use the diagram to name the figures.

4. Three noncollinear points
5. Three collinear points
6. Four noncoplanar points
7. Two intersecting lines

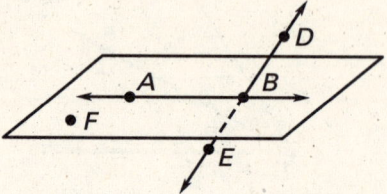

Find the length of the segment.

8. $\overline{AD}$
9. $\overline{CB}$

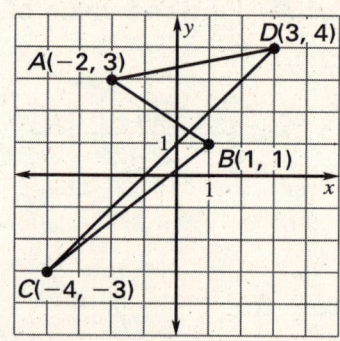

Use the diagram to find the measure of the angle.

10. $\angle AMC$
11. $\angle CMV$
12. $\angle AMP$

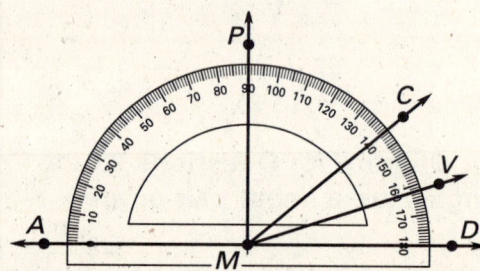

In Questions 13–15, refer to the diagram for Questions 10–12. State what type of angle is formed.

13. $\angle AMC$
14. $\angle CMV$
15. $\angle AMP$

Find the coordinates of the midpoint of a segment with the given endpoints.

16. $A(0, -4)$
    $B(4, 0)$
17. $C(3, 10)$
    $D(-1, -6)$
18. $E(4, -8)$
    $F(-8, -4)$

Answers

1. _____
2. _____
3. _____
4. _____
5. _____
6. _____
7. _____
8. _____
9. _____
10. _____
11. _____
12. _____
13. _____
14. _____
15. _____
16. _____
17. _____
18. _____

Chapter Test C

For use after Chapter 1

In Questions 19 and 20, $\vec{EF}$ is the angle bisector of $\angle TEA$. Find the two angle measures not given in the diagram.

19.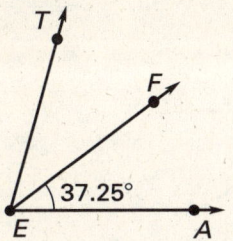

20.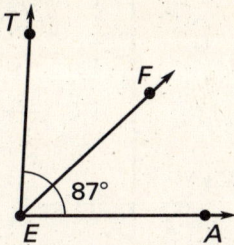

In Questions 21–22, $\vec{BD}$ bisects $\angle ABC$. Find the value of x.

21.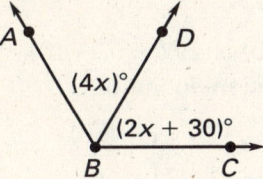

22.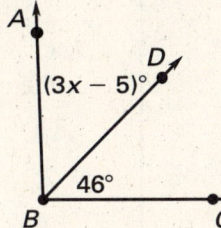

In Questions 23–25, use the diagram to solve for the missing angle measure.

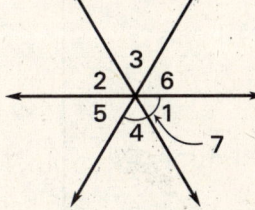

23. If $m\angle 1 = 37°$, then $m\angle 2 = $ ____.

24. If $m\angle 4 = 72°$ and $m\angle 5 = 43°$, then $m\angle 1 = $ ____.

25. If $m\angle 2 = m\angle 6 = x°$, then $m\angle 3 = $ ____.

In Questions 26–28, refer to the diagram for Questions 23–25 and state whether the given angles are a linear pair, vertical angles, or neither.

26. $\angle 1$ and $\angle 2$ 27. $\angle 2$ and $\angle 3$ 28. $\angle 5$ and $\angle 7$

Find the values of the variables.

29.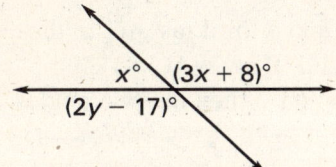

30.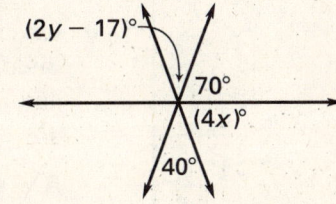

Find the area of the figure described.

31. Triangle with height 3.25 cm and base of 5 cm

32. Square with side length $4\frac{1}{2}$ ft

33. Circle with radius 6 yd

Answers

19. _____
20. _____
21. _____
22. _____
23. _____
24. _____
25. _____
26. _____
27. _____
28. _____
29. _____
30. _____
31. _____
32. _____
33. _____

CHAPTER 1

SAT/ACT Chapter Test

For use after Chapter 1

1. What is the next number in the sequence
   $-3, -\frac{7}{2}, -4, -\frac{9}{2}, ...?$
   - (A) -3
   - (B) -5
   - (C) $-\frac{11}{2}$
   - (D) -7
   - (E) $-\frac{16}{2}$

2. Which statement(s) may be true about the two lines shown in the diagram?

   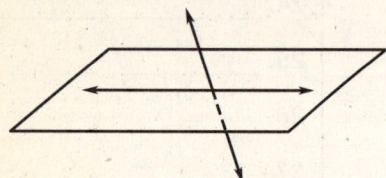

   I. The lines are coplanar.
   II. The lines are not coplanar.
   III. The lines intersect in one point.

   - (A) I only
   - (B) I and II only
   - (C) II and III only
   - (D) I and III only
   - (E) I, II, and III

3. What is the distance between point $A(-3, 2)$ and point $B(5, -1)$?
   - (A) $\sqrt{5}$
   - (B) 73
   - (C) $\sqrt{73}$
   - (D) 11
   - (E) 5

4. In the diagram, what are the values of x and y?

   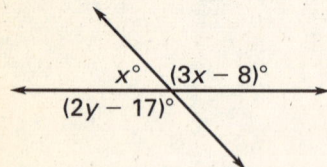

   - (A) $x = 47, y = 75$
   - (B) $x = 47, y = 74$
   - (C) $x = 75, y = 47$
   - (D) $x = 71, y = 51$
   - (E) $x = 45, y = 77$

5. $\angle A$ and $\angle T$ are complementary. The measure of $\angle T$ is three times the measure of $\angle A$. What is $m\angle A$?
   - (A) $22°$
   - (B) $22.5°$
   - (C) $23°$
   - (D) $24°$
   - (E) $23.5°$

6. $\angle 1$ and $\angle 2$ in the diagram are ____?

   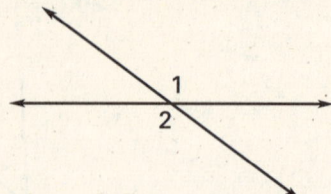

   - (A) vertical angles
   - (B) complementary
   - (C) a linear pair
   - (D) supplementary

7. Given points $G(2, 10)$ and $H(-6, -10)$, find the coordinates of the midpoint of $\overline{GH}$.
   - (A) $(-2, 10)$
   - (B) $(-4, 0)$
   - (C) $(-2, 0)$
   - (D) $(8, 20)$
   - (E) $(-4, -10)$

8. Which of the following statements is *false*?

   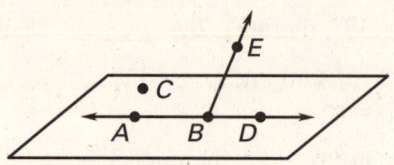

   - (A) A, B, C, and D are coplanar.
   - (B) A, B, and D are collinear.
   - (C) $\overrightarrow{BE}$ and $\overrightarrow{BA}$ are opposite rays.
   - (D) Answers B and C only
   - (E) Answers A, B, and C

9. Given $\angle BAD$, and a third ray AC in the interior of $\angle BAD$, if $m\angle BAC = 129°$ and $m\angle CAD = 51°$, then the two angles are ____?
   - (A) supplementary
   - (B) complementary
   - (C) a linear pair
   - (D) supplementary and a linear pair
   - (E) complementary and a linear pair

116 Geometry
Chapter 1 Resource Book

CHAPTER 1
Alternative Assessment and Math Journal
For use after Chapter 1

JOURNAL 1. Draw three noncollinear points. Label them *E*, *F*, and *G*. Draw line *AE* so that *F* is between *A* and *E* and line *BG* so that *F* is between *B* and *G*. Name two opposite rays. Name one pair of vertical angles. Name one pair of supplementary angles. Find the sum of the measures of ∠AFB and ∠EFB.

MULTI-STEP PROBLEM 2. The diagram below shows the location of eight marbles. ∠TAP is a right angle. ∠KAL is a right angle. The measure of ∠TAN is 14°. Marble *A* is at (3, 0). Marbles *S*, *A*, and *P* are collinear.

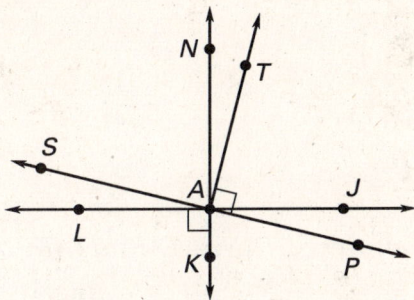

a. Marble *K* is at (3, −6) and Marble *N* is at (3, 20). Find the distance between *K* and *N*. Find the coordinates of the midpoint between the two marbles and label this point *M*.

b. Find the length of $\overline{AM}$.

c. Find *m*∠SAN.

d. Find *m*∠NAL.

e. Name an acute angle, a right angle, an obtuse angle, and a straight angle.

f. Imagine ray $\overrightarrow{AZ}$ bisects ∠TAP. Find *m*∠TAZ.

3. *Critical Thinking* In the diagram above, add the necessary points or lines to create the figure.

a. Find the perimeter of the triangle defined by the three marbles *L*(−7, 0), *A*(3, 0), and *K*(3, −6).

b. Add a marble at point *Q* with coordinates of (−10, 16). The four marbles *T*(6, 12), *A*(3, 0), *S*(−13, 4), and *Q* form a rectangle. Find the perimeter and area of the rectangle.

4. *Writing* There are four ways to classify a single angle according to its measure. Name each classification. Write a sentence describing the angle measure in each classification. Draw a detailed diagram to represent each classification.

CHAPTER 1 CONTINUED

Alternative Assessment Rubric
For use after Chapter 1

JOURNAL SOLUTION

1. Complete answers should include:
   - A diagram with E, F, and G not collinear.
   - Also, $\overleftrightarrow{AE}$ intersecting $\overleftrightarrow{BG}$ at the point F.
   - Two opposite rays: $\overrightarrow{FB}$ and $\overrightarrow{FG}$ or $\overrightarrow{FE}$ and $\overrightarrow{FA}$.
   - Vertical angles: $\angle BFA$ and $\angle GFE$ or $\angle BFE$ and $\angle GFA$.
   - Supplementary angles: $\angle BFA$ and $\angle AFG$, or $\angle GFE$ and $\angle EFB$, or $\angle AFG$ and $\angle GFE$, or $\angle EFB$ and $\angle BFA$.
   - $m\angle AFB + m\angle EFB = 180°$

MULTI-STEP PROBLEM SOLUTION

2. **a.** 26 units; $M(3, 7)$

   b. 7 units

   c. 76°

   d. 90°

   e. Answers may vary. *Sample answer:* $\angle TAN$; $\angle TAP$; $\angle TAK$; $\angle SAP$

   f. 45°

3. **a.** About 27.7 units

   b. About 57.7 units and 204 square units

4. Answers may vary. Answers should include acute angle, right angle, obtuse angle, and straight angle. Answers should have a diagram and an explanation of each classification.

MULTI-STEP PROBLEM RUBRIC

4 Students answer all parts of the problem correctly, showing all work. Students add correct points and lines to the diagram. Students name and explain all four angle classifications very clearly.

3 Students complete all parts of the problem. Students' work may have small mathematical errors. Students' diagrams may have minor errors. Students name and explain all four angle classifications.

2 Students complete all parts of the problem. Students' work may have several mathematical errors. Students' diagrams are incorrect. Students' explanations are not clear and do not match the angle.

1 Students do not complete the problem. Students' diagrams are incomplete. Students do not name and explain four angle classifications.

CHAPTER 1

NAME _____ DATE _____

Project: Creating Paper Measuring Tools

For use with Chapter 1

OBJECTIVE Explore the measures of angles and their relationships.

MATERIALS A few square pieces of paper

INVESTIGATION Carry out the following set of steps two times. In the exercises you will label the squares you folded to create your own measuring tools.

> *Step 1*: Start with a square piece of paper. Fold the paper in half horizontally and crease it. Then open it up.

Step 2: Carefully fold the top right vertex down to the horizontal crease line so that the new crease passes through the upper left vertex.	*Step 3*: Fold the bottom left vertex up to meet the horizontal crease line so that this new crease line also passes through the upper left vertex.	*Step 4*: Fold along the edge created by the two folded triangles. The lower right vertex should come up and over the triangles as in the diagram.
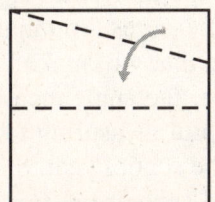	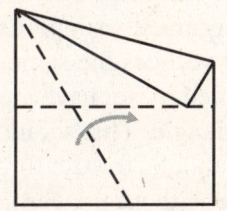	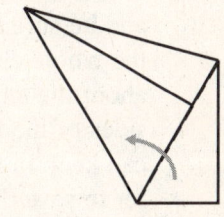

Open the paper to examine the shapes and angles created by the fold lines.

1. On one of your folded squares, label the vertices of the square and the endpoints and intersections of the fold lines. Identify the shapes in the seven regions formed.

2. Use the types of angles and angle relationships you have learned about in the chapter to find the measures of as many of the angles formed by the folds as you can without using a protractor. (For example, if two angles are congruent and the sum of their angle measures is 90°, then the measure of each angle is 45°.) Label the measures you identify on both squares. Keep track of your reasoning since this will be part of your project report.

3. You can use angles whose measures you already know to "measure" other angles. For example, you can place a 45°-angle from one paper square against an angle with unknown measure on the other paper square. If the angles are congruent, then the unknown angle also measures 45°. Use this approach and any angle relationships you observe to find the missing angle measures for the angles formed by the folds. Label both squares.

PRESENT YOUR RESULTS Write a report that explains how you determined each angle measure for your paper tool. Include a drawing of the open square with the vertices labeled and the angle measures written in. How could you use this tool?

Project: Teacher's Notes
For use with Chapter 1

GOALS
- Classify and measure angles.
- Identify angle relationships (such as vertical angles, complementary angles, and supplementary angles) and apply their properties.
- Identify polygons.

MANAGING THE PROJECT

Classroom Management Consider using 8.5 in. × 8.5 in. squares of paper. The larger the square of paper, the easier it is for students to label and write in the angle measures. Students may find it easier to get crisp fold lines using fairly thin paper or even tracing paper. You may wish to have students work in pairs. If so, they only need to fold one square apiece since partners can "measure" with each other's square to complete their labeling. Using a protractor to measure the angles should be a last resort, but could be used for checking final results. Every fold will be a little different with some degree of error. You want each student to end up with the same measures for their angles.

Guiding Students' Work Students need to fold their paper accurately and neatly. Students should share insights as they determine the individual angle measures and be sure to give arguments supporting those measures. You may wish to have the whole class label vertices the same way since that will make it easier to talk about the angles formed. Also, have students write the angle measures on both sides of the folded triangle. This should be a helpful measuring tool that students can keep in their notebooks for use throughout the school year. Students may prefer to save the square without the vertices labeled as their reference tool since the labeling may be less confusing.

RUBRIC The following rubric can be used to assess student work.

4 The student accurately draws and labels a sketch of the folded square. The student writes a clear description, including calculations, of how he or she found each angle measure on the folded square of paper. The student writes a thoughtful assessment of the value of the measuring tool.

3 The student draws and labels a fairly accurate sketch of the folded square. The student writes a description, including calculations, of how she or he found each angle measure, but there are minor errors in the calculations or the description could be clearer. The student writes an assessment of the value of the measuring tool.

2 The student draws and labels a sketch of the folded square, but it may be messy or some angle measures may be missing. The descriptions and calculations showing how he or she found each angle measure contain several errors. The student writes a poor assessment of the value of the measuring tool.

1 The sketch of the folded square is inaccurate or incomplete. Descriptions and calculations for angle measures are missing or contain many errors. The assessment of the value of the measuring tool is missing or very poorly written.

CHAPTER 1

NAME _____ DATE _____

Cumulative Review

For use after Chapter 1

Describe a pattern in the sequence of numbers. Predict the next number (1.1)

1. 2, 6, 14, 30, ...
2. 1, 4, 9, 16, ...
3. 96, 48, 24, 12, ...
4. 3125, 625, 125, 25, ...

Complete the conjecture based on the pattern you observe in these specific cases. (1.1)

5. Conjecture: The product of two consecutive positive integers is always ? .

   $3 \times 4 = 12 \qquad 7 \times 8 = 56$
   $4 \times 5 = 20 \qquad 10 \times 11 = 110$

6. Conjecture: The square of any odd integer is always ? .

   $7^2 = 49 \qquad 11^2 = 121 \qquad 13^2 = 169$
   $9^2 = 81 \qquad 15^2 = 225 \qquad 17^2 = 289$

Decide whether the statement is *true* or *false*. (1.2)

7. Point C lies on line l.
8. Point E lies on $\overline{AB}$.
9. Points D, A, and B are collinear.
10. Points D, A, and B are coplanar.
11. Point C lies on line m.
12. Lines l and m intersect at E.

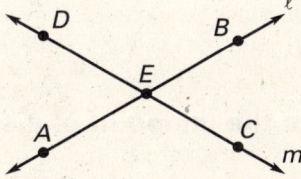

Find the length of each segment. (1.3)

13. $AD = 30$
    $AB = 2x + 2$
    $BC = 4x - 1$
    $CD = 3x - 7$

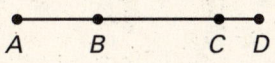

Find the distance between each pair of points. Round your answers to the nearest hundreth. (1.3)

14. $MN = $?
15. $NO = $?
16. $OP = $?
17. $PM = $?
18. $MO = $?
19. $NP = $?

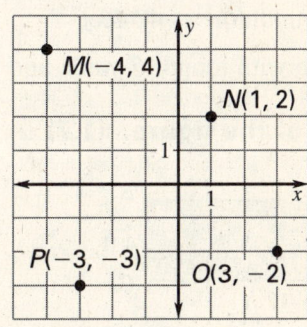

CHAPTER 1 CONTINUED

NAME _____ DATE _____

Cumulative Review

For use after Chapter 1

Use a protractor to draw the angle described. (1.4)

20. $m\angle ABC = 40°$

21. $m\angle DEF = 120°$

Use a protractor to measure the angle to the nearest degree. (1.4)

22.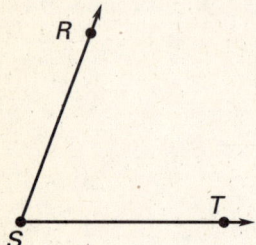

23.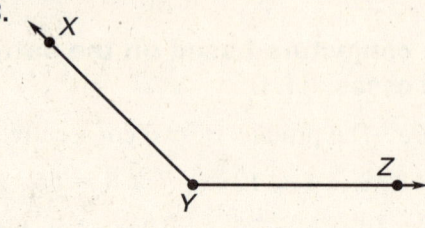

Use the Angle Addition Postulate to find the measure of the unknown angle. (1.4)

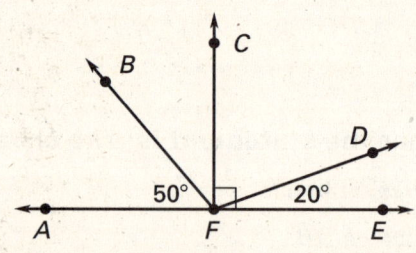

24. $m\angle CFD = \underline{\ ?\ }$

25. $m\angle BFC = \underline{\ ?\ }$

26. $m\angle AFD = \underline{\ ?\ }$

27. $m\angle BFE = \underline{\ ?\ }$

Find the coordinates of the midpoint of a segment with the given endpoints. (1.5)

28. $A(7, 3), B(9, -1)$

29. $A(-6, 6), B(4, 10)$

30. $A(-2, -6), B(7, 0)$

31. $A(12, 5), B(3, -3)$

$\angle A$ and $\angle B$ are complementary. Find $m\angle A$ and $m\angle B$. (1.6)

32. $m\angle A = 7x + 1$
    $m\angle B = 5x - 7$

33. $m\angle A = 5x + 11$
    $m\angle B = 2x - 5$

Find the perimeter (or circumference) of the figured described. (When necessary use $\pi \approx 3.14$.) (1.7)

34. Circle with diameter 40 feet

35. Rectangle with length 6 yards and width 3 yards

Find the area of the figure. (1.7)

36.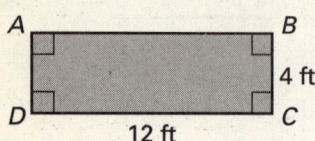

37.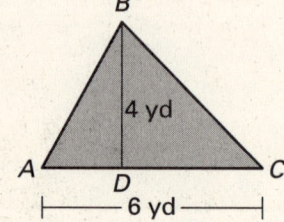

122   Geometry
      Chapter 1 Resource Book

ANSWERS

Chapter Support

Parent Guide
1.1: Each number after the first two is the sum of the two numbers before it. The next number is 52.
1.2: *Sample answer:* where the floor and one of the walls meet **1.3:** Yes, both have length $\sqrt{29}$.
1.4: acute; obtuse **1.5:** $(17, -7)$
1.6: 60° and 120° **1.7:** $226.08

Prerequisite Skills Review
1. 13 **2.** -14 **3.** 17 **4.** -1 **5.** 7 **6.** -9
7. 40 **8.** 41 **9.** 41 **10.** 73 **11.** 36 **12.** 82
13. 3.61 **14.** 7.81 **15.** 15.81 **16.** 10.63
17. 12.65 **18.** 14.87

Strategies for Reading Mathematics
1. *Sample answer:* I should understand and use the basic undefined terms and defined terms of geometry, and sketch the intersections of lines and planes. **2.** in the "Why you should learn it" statement in the left margin; perspective drawing
3. study tips and skills reviews; study tips help you avoid common errors; skills reviews direct you to pages that explain concepts from earlier math classes.

Lesson 1.1

Warm-Up Exercises
1. 0.25, 1, 1 **2.** 0, 3, 8 **3.** 1, $\frac{3}{2}$, 2 **4.** $\frac{1}{2}$, 2, 3

Lesson Opener
Allow 10 minutes.
1. 81; 9801; 998,001 **2.** 99,980,001; 9,999,800,001 **3.** 111,105; 222,210; 333,315
4. 444,420; 555,525 **5.** 9; 1089; 110,889
6. 11,108,889; 1,111,088,889

Practice A
1. **2.** 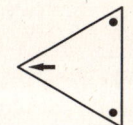 **3.**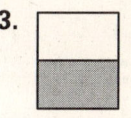

4.

In answers 5–12, the description of the pattern may vary. A sample answer is given.
5. adding 3 to each term; 14 **6.** dividing each term by 3; $\frac{1}{3}$ **7.** add 111 to each term; 567
8. add consecutive even numbers; 35
9. add 1 to numerator and 1 to denominator; $\frac{5}{6}$
10. subtract 1 from numerator, add 2 to denominator; $\frac{1}{12}$
11. subtract 3 from each term; -8
12. square numbers; 25 **13.** 13 **14.** 15
15. an even number **16.** an odd number

Practice B
1. **2.**  **3.**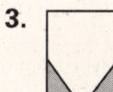

4.

In answers 5–12, the description of the pattern may vary. A sample answer is given.
5. add 111 to each term; 557 **6.** add consecutive integers; 25 **7.** add 2 to numerator, add 1 to denominator; $\frac{9}{6}$ **8.** subtract 1 from numerator and from denominator; $\frac{1}{2}$ **9.** subtract 4 from each term; -12 **10.** square numbers; 36
11. *Sample:* double the number and add 1; 47
12. Add consecutive multiples of 3; 47
13. 16 **14.** 20
15. Answers may vary. *Sample:* $2 \div 4 = 0.5$

Lesson 1.1 continued

16. Answers may vary. Sample: $|3| - |4| = -1$

17. Answers may vary. Sample: $\dfrac{-2}{-2+1} = 2$

Practice C

1.   2. 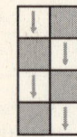 3.

4.

In answers 5–12, the description of the pattern may vary. A sample answer is given.

5. add 10, subtract 20, add 10, subtract 20; 103

6. add consecutive odd numbers; 40

7. subtract consecutive whole numbers; 8

8. square consecutive tenths; 1.21 **9.** Sample: double the number and subtract 1; 113

10. add 3 to each term; 5

11. multiply each term by -2; 32 **12.** add 3 in numerator, subtract 2 in the denominator; $\dfrac{15}{-1}$ (or -15) **13.** 384 **14.** 5 **15.** Answers may vary. Sample: $5 - 6 = -1$ **16.** Answers may vary. Sample: $|-2 + 3| \neq |-2| + |3|$

17. Answers may vary. Sample: $\dfrac{-2}{-2-1} = \dfrac{2}{3}$

Reteaching with Practice

1.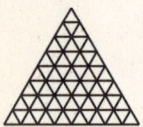

2. The second number is 2 times the first, the third number is 3 times the second, and so on; 120

3. *Sample:* The first number is 1 squared then minus 1, the second number is 2 squared then minus 1, the third number is 3 squared then minus 1, and so on; 35 **4.** $a^2 - b^2$

5. *Sample:* $(4 + 5)^2 = 81 \neq 41 = 4^2 + 5^2$

Real-Life Application

1. a. 11 **b.** 5 **2.** 55 **3.** 10 pairs of numbers that sum to 21; 210 **4.** The sum of each pair will be the same as the sum of the first and last number. The total number of pairs will be half the last number. Multiply the sum by the number of pairs. **5.** 5050 **6.** The sum of each pair will be computed the same as before. The number of pairs will be $\frac{1}{4}$ of the last number.

Challenge: Skills and Applications

1. The numbers are the squares of odd numbers; 81, 121, 169 **2.** 1 **3.** If the square of an odd number is divided by 4, the remainder is 1.

4. Sample answer: $(2n + 1)^2 = 4n^2 + 4n + 1 = 4n(n + 1) + 1$, so if $(2n + 1)^2$ is divided by 4, the quotient is $n(n + 1)$ and the remainder is 1. **5.** Sample answer: The nth number is n more than the previous number; 15, 21, 28, 36, 45.

6. $T_n = \dfrac{n(n+1)}{2}$ **7.** 1, 2, 4, 7

8. Sample answer: You would sketch the fourth chord so that it intersects each of the existing chords in a different location; 4 additional regions.

9. n **10.** For 0 chords, there is 1 region. For n chords ($n \geq 1$), the number of regions is $1 + (\underbrace{1 + 2 + 3 + 4 + \cdots + n})$.

sum of first n integers

11. $R_n = \frac{1}{2}n^2 + \frac{1}{2}n + 1$

Lesson 1.2

Warm-Up Exercises

1. $(-2, 1)$ **2.** $(4, 0)$ **3.** $(3, -2)$
4. $(-1, -3)$

Daily Homework Quiz

1. double each number and add 1; 191

2. triple each number and subtract 20; -557

3. 2 and 3 are both prime and are consecutive.

4. Zero is an integer but division by zero is not defined.

Lesson 1.2 continued

Lesson Opener
Allow 10 minutes.

1. Check drawing. There should be a dot at the intersection of Oak and Pine.   2. Check drawing. There should be a dot at the intersection of Oak and Maple.   3. Check drawing. There should be a dot at the intersection of Pine and High.

4. They appear to be on the same line.

5. Check drawing. There should be a dot at the intersection of Main and Front.   6. Check drawing. There should be a dot at the intersection of Main and a street other than Front.   7. Check drawing. There should be a dot at the intersection of Front and a street other than Main.

8. They do not appear to be on the same line.

Practice A
1–4. Sample figures are given.

1.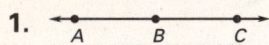

2. 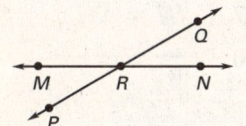   3.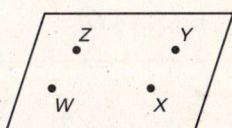

4.   5. true   6. true

7. false   8. true   9. true   10. false   11. true
12. true   13. H   14. G   15. A   16. B   17. B
18. E   19. G   20. A   21. P   22. Q   23. P
24. S   25. Q   26. P   27. N   28. N

29. lie on the same line   30. lie in the same plane   31. between X and Y   32. on the same side of M as point N   33. lie in the same line
34. P, Q, and T are collinear and P is between Q and T   35. $\vec{BC}$ has initial point B and extends in the direction of C. $\vec{CB}$ has initial point C and extends in the direction of B.

Practice B
1. true   2. false   3. false   4. true   5. true
6. true   7. false   8. true   9. I   10. E
11. B   12. B   13. C   14. D   15. E   16. A
17. T   18. P   19. P   20. S   21. Q   22. P
23. N   24. N   25. between A and B

26. on the same side of P as point Q
27. lie in the same line   28. M, N, and L are collinear and M is between N and L

29–36. Sample figures are given.

29.   30.

31. 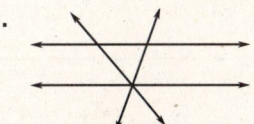   32.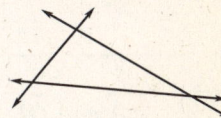

33. 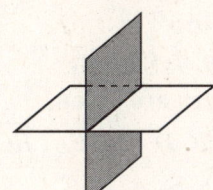   34.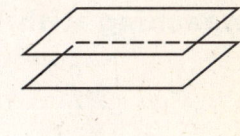

35.    36.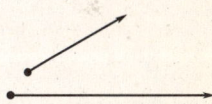

Practice C
1. true   2. true   3. false   4. true   5. false
6. true   7. true   8. true   9. A   10. G
11. A   12. F   13. F   14. A   15. G   16. D
17. D   18. A   19. J   20. B   21. E   22. J
23. C   24. F

25–36. Sample figures are given.

25. 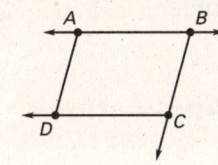   26.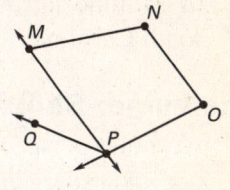

27.   28.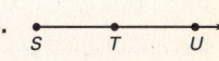

29. not possible   30.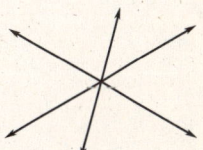

Lesson 1.2 continued

31. 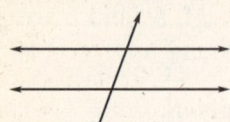 32.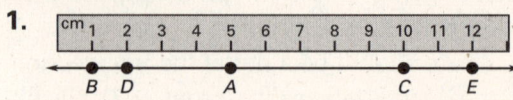

33.  34. not possible

35. 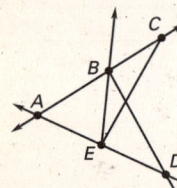 36.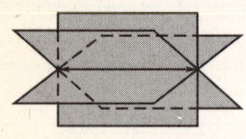
$\overrightarrow{PQ}$ and $\overrightarrow{QP}$

Reteaching with Practice

1. 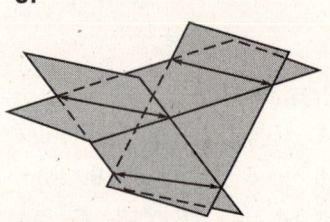   2, 4, 6, 7. Sample answers are given.
2. A, B, D 3. $\overrightarrow{BA}, \overrightarrow{BC}$
4. $\overleftrightarrow{BD}$ 5. A, B, C
6. $\overrightarrow{BA}, \overrightarrow{BD}$ 7. $\overleftrightarrow{AB}, \overleftrightarrow{BC}$

8.   9.

Interdisciplinary Application

1. Cleveland, OH; Columbia, SC; Jacksonville, FL; Key West, FL 2. Los Angeles, CA
3. 38° latitude and 77° longitude
4. South Carolina

Challenge: Skills and Applications

1. two-point perspective
2. one-point perspective 3. neither
4. two-point perspective 5. Check drawings.

Lesson 1.3

Warm-Up Exercises
1. 3 2. 3.1 3. 5 4. $\sqrt{13}$

Daily Homework Quiz
1. true 2. true 3. false 4. false 5. true

Lesson Opener
Allow 15 minutes.
1.

2. 3 cm 3. 4 cm 4. 5 cm 5. 10 cm
6. 8 cm 7. 9 cm.

Practice A
1. 32 mm 2. 24 mm 3. 29 mm 4. 34 mm
5. 21 mm 6. 14 mm
7. $DS + SP = DP$ 8. $SJ + JH = SH$
9. $QC + CR = QR$ 10. $MT + TN = MN$
11. 5 12. 5 13. 5 14. 9 15. 10 16. 14
17. $x = 8$; $HJ = 40$, $JK = 56$ 18. $x = 4$; $HJ = 13$, $JK = 5$ 19. $x = 14$; $HJ = 79$, $JK = 50$ 20. $\sqrt{5}$ 21. $\sqrt{10}$ 22. 5
23. congruent 24. not congruent
25. congruent

Practice B
1. 35 mm 2. 18 mm 3. 32 mm
4. $TA + AQ = TQ$ 5. $HM + MA = HA$
6. $SJ + JH = SH$ 7. $LA + AB = LB$

Sketch for Exercises 8–11:

8. $RS = 4$ 9. $QS = 14$ 10. $TS = 8$
11. $TV = 12$ 12. $x = 3$; $HJ = 10$, $JK = 12$
13. $x = 11$; $HJ = 52$, $JK = 79$

Lesson 1.3 continued

14. $x = 1$; $HJ = 2\frac{1}{3}$, $JK = 5\frac{2}{3}$, $KH = 8$
15. $DE = \sqrt{10}$, $EF = 2\sqrt{17}$, $FD = 5\sqrt{2}$
16. $GH = 5$, $HI = 5\sqrt{2}$, $IG = \sqrt{13}$
17. $AB = \sqrt{5}$, $BC = \sqrt{10}$, $CA = \sqrt{29}$
18. b; distance $\approx$ 26.4 miles

Practice C

1. [diagram: T D Q]
   $TD + DQ = TQ$
2. [diagram: Q M N]
   $QM + MN = QN$
3. [diagram: T L W]
   $TL + LW = TW$
4. [diagram: X A Y]
   $XA + AY = XY$

Sketch for Exercises 5–8:
[diagram with segments 8 and 23, points Q T R S V]

5. $RS = 5$ 6. $QS = 18$ 7. $TS = 10$
8. $TV = 15$ 9. $x = 7$; $HJ = 27$, $JK = 17$
10. $x = 6$; $HJ = 45$, $JK = 67$
11. $x = 11$; $HJ = 7\frac{2}{3}$, $JK = 22\frac{2}{3}$, $HK = 30\frac{1}{3}$
12. $DE = \sqrt{5}$, $EF = 2\sqrt{37}$, $FD = \sqrt{101}$
13. $GH = 5\sqrt{5}$, $HI = \sqrt{73}$, $IG = 10$
14. $AB = \sqrt{13}$, $BC = \sqrt{41}$, $CA = \sqrt{58}$
15. congruent   16. not congruent
17. congruent   18. b; distance $\approx$ 26.4 miles

Reteaching with Practice

1. a. 10 b. 5 c. $\overline{AB}$ and $\overline{BC}$ are congruent.
2. a. 4 b. 4 c. 5 d. 5   3. $7\sqrt{5}$
4. $\sqrt{61}$ 5. $\sqrt{1261}$ 6. $4\sqrt{10}$ 7. $\sqrt{89}$
8. $2\sqrt{a^2 + b^2}$

Cooperative Learning Activity

1. *Sample answer*: No. The Distance Formula was not needed if the segment connecting two points was horizontal or vertical. Counting or subtraction easily determined the distance.
2. *Sample answer*: Estimation and comparison determined the distances to try first.

Real-Life Application

1.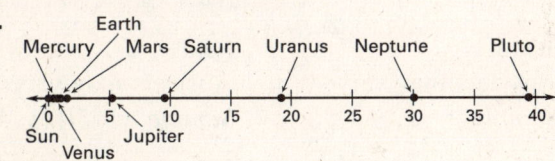

2. Mercury and Venus 3. Venus 4. 0.52 AU
5. Uranus and Neptune 6. About 0.003 AU
7. The moon would be at 0.997 AU during a solar eclipse and at 1.003 AU during a lunar eclipse.

Math and History

1. [segment AB]

2.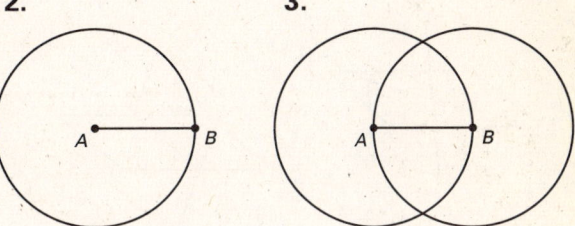

3.

4.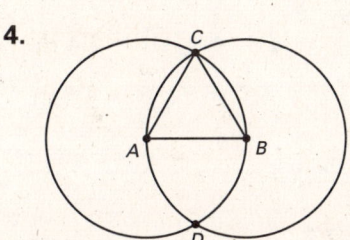

5. Answers may vary. 6. Repeat Exercises 1–4. Use a straightedge to draw $\overline{CD}$. Let P be the point where $\overline{CD}$ and $\overline{AB}$ intersect. $\triangle APC$ and $\triangle BPC$ are right triangles.

Challenge: Skills and Applications

1. *Sample answer*:
$LR = LM + MR = RS + MR = MS$
2. $\overline{UW} \cong \overline{VX} \cong \overline{WY} \cong \overline{XZ}$, $\overline{UX} \cong \overline{WZ}$, $\overline{UY} \cong \overline{VZ}$
3. *Sample answer*: The first half of the Segment Addition Postulate requires B to be between A and C, not just on the same line.
4. *Sample answer*: [diagram: A C B]
5. no; *Sample answer*: [diagram: A B D C]
6. yes; *Sample answer*: $AB + BC + CD$
   $= AB + (BC + CD) = AB + BD = AD$
7. no; *Sample answer*: [diagram: A C B D]
8. 8 9. -4 10. 126 11. no 12. yes
13. no

Lesson 1.3 continued

Quiz 1
1. 14 2. 9

3–6. Sample figures are given.

3.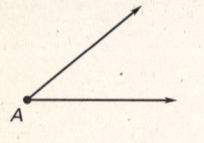

4.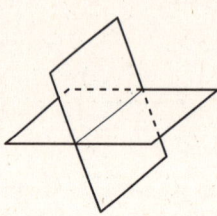

5.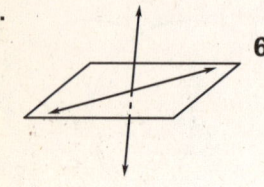

6.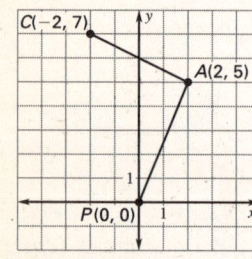

7. 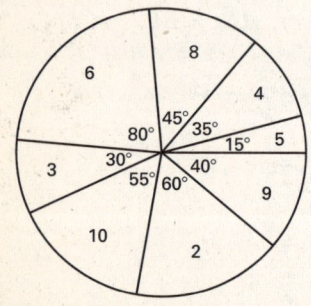 $PA = \sqrt{29} \approx 5.4$ m
$AC = 2\sqrt{5} \approx 4.5$ m

Lesson 1.4

Warm-Up Exercises
1. R 2. congruent 3. postulate or axiom
4. line

Daily Homework Quiz
1. 7 2. 8 3. no 4. yes

Lesson Opener
Allow 15 minutes.

1. piece 1; *Sample answer:* no, that piece is so large that there wouldn't be room for seven of the other pieces within the circle.

2. (Angle measures are shown for your reference. Note that the angle measure of piece 1 is 510° and the angle measure of piece 7 is 70°.)

3. *Sample answer:* If the largest piece (piece 1) is in the pan, eight pieces will not fit. If the two next-to-largest pieces (piece 6 and piece 7) are both in the pan, eight pieces will not fit, so one of them must be left out along with piece 1. If pieces 6 and 1 are left out, the pieces do not quite fill a circle, but leaving out pieces 7 and 1 works.

Practice A
1. Q; $\vec{QP}$, $\vec{QT}$; ∠PQT, ∠TQP, ∠Q
2. A; $\vec{AR}$, $\vec{AM}$; ∠RAM, ∠MAR, ∠A
3. F; $\vec{FJ}$, $\vec{FK}$; ∠JFK, ∠KFJ, ∠F   4. 125°
5. 21° 6. 90° 7. 70° 8. 180° 9. 133°
10. obtuse; 135° 11. right; 90° 12. acute; 45°

13.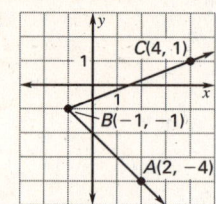
acute;
Answers may vary;
Sample: (1, −1); (0, 1).

14.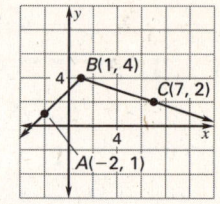
obtuse;
Answers may vary;
Sample: (1, 3); (1, 5).

15.
right; Answers may vary;
Sample: (2, −1); (2, −3).

Practice B
1. 133°; ∠CDP, ∠PDC, ∠D
2. 27°; ∠FYS, ∠SYF, ∠Y
3. 65°; ∠ABC, ∠CAB, ∠B
4. 90° 5. 142° 6. 53°

7.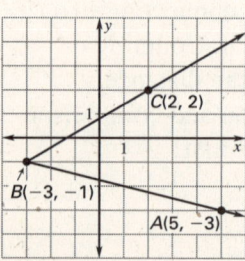
acute;
Answers may vary;
Sample: (−2, −1); (−4, −1)

8.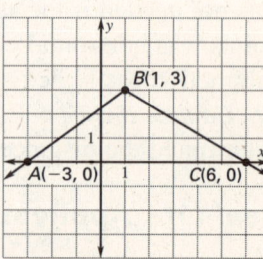
obtuse;
Answers may vary;
Sample: (1, 2); (1, 4).

Lesson 1.4 continued

9. 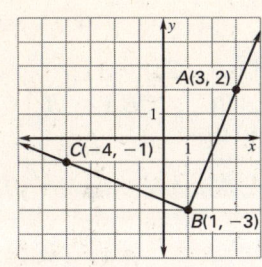 right; Answers may vary. *Sample*: (1, −2); (2, −4)

Sketch for Exercises 10–13:

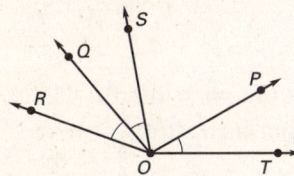

10. $m\angle QOP = 100°$ **11.** $m\angle QOT = 130°$
12. $m\angle ROQ = 30°$ **13.** $m\angle SOP = 70°$

Diagram for Exercises 14–17:

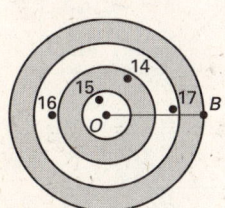

14. 50
15. 100
16. 25
17. 25
18. 200

Practice C

1. 67°; $\angle JKL$, $\angle LKJ$, $\angle K$
2. 121°; $\angle TMQ$, $\angle QMT$, $\angle M$
3. 34°; $\angle ACU$, $\angle UCA$, $\angle C$ **4.** 44° **5.** 130°
6. 57°

7. 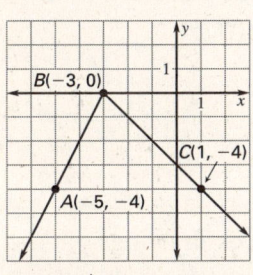 acute; Answers may vary. *Sample*: (−3, −1); (−3, 2)

8. 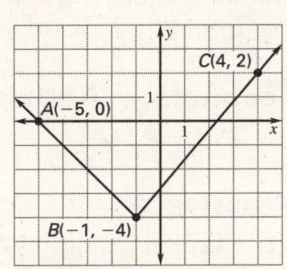 acute; Answers may vary. *Sample*: (−1, −3); (−1, −5)

9. 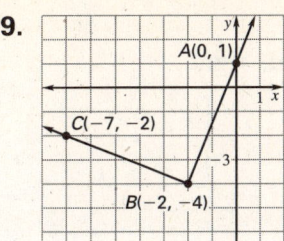 right; Answers may vary. *Sample*: (−2, −3); (−2, −5)

Sketch for Exercises 10–13:

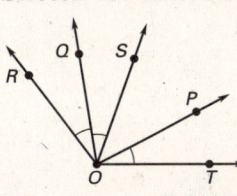

10. $m\angle QOP = 71°$
11. $m\angle QOT = 99°$
12. $m\angle ROQ = 28°$
13. $m\angle SOP = 43°$
14. $x = 8$; $m\angle POQ = 12°$, $m\angle QOR = 14°$
15. $x = 7$; $m\angle POQ = 28°$, $m\angle QOR = 33°$
16. $x = 1$; $m\angle POQ = \frac{2°}{3}$, $m\angle QOR = 3\frac{1°}{3}$, $m\angle POR = 4°$

Reteaching with Practice

1. $\angle ABC$, $\angle CBA$, $\angle B$; vertex B, sides $\overrightarrow{BA}$ and $\overrightarrow{BC}$ **2.** $\angle DEF$, $\angle FED$, $\angle E$; vertex E, sides $\overrightarrow{EF}$ and $\overrightarrow{ED}$ **3.** 35°

4. a. acute **b.** obtuse

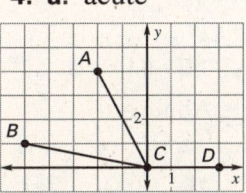 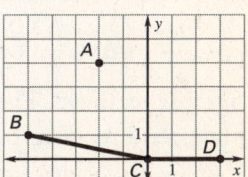

c. obtuse

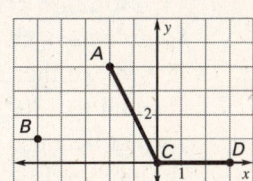

5. a. straight **b.** acute

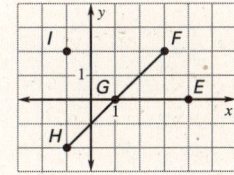 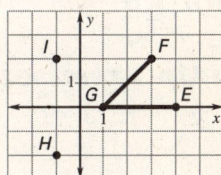

Geometry
Chapter 1 Resource Book

Lesson 1.4 continued

c. obtuse
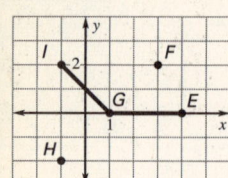

d. right
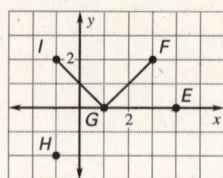

e. right
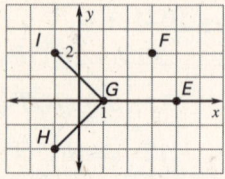

Interdisciplinary Application
1. ∠PRI 2. 150° 3. 120° 4. 10° 5. about 87%; the ability to drive a car would be impaired by the lack of peripheral vision. Such a narrow field of vision would affect safety at intersections, railroad crossings, and so on.

Challenge: Skills and Applications
1. *Sample answer:*
Since ∠1 ≅ ∠3, m∠1 = m∠3. Therefore, m∠BAD = m∠1 + m∠CAD = m∠3 + m∠CAD = m∠CAE. So, ∠BAD ≅ ∠CAE.
2. ∠1 ≅ ∠5 (or ∠RXS ≅ ∠VXW), ∠SXU ≅ ∠TXV, ∠RXU ≅ ∠TXW, ∠RXV ≅ ∠SXW 3. 15 4. 7 5. 84° 6. 54°
7. *Sample answer:* The Angle Addition Postulate requires P to be in the interior of ∠RST.
8. *Sample answer:*
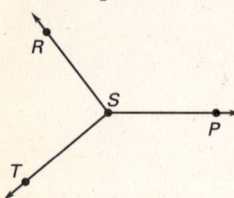
9. $2x°$ 10. $(360 - 2x)°$
11. Both are correct; *Sample answer:* Since $x = 90°$, either expression is equal to 180°, which is correct because ∠RST is a straight angle.

Lesson 1.5

Warm-Up Exercises
1. 5 2. −1 3. 50 4. 17.5

Daily Homework Quiz
Graph for Exercises 1–4:
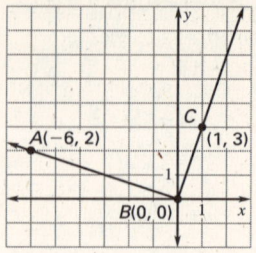
1. B 2. two of ∠ABC, ∠CBA, ∠B
3. 90°
4. *Sample answer:* (0, 2)

Lesson Opener
Allow 10 minutes.
1. *Sample answer:* It cuts the park in half; Check drawings: sidewalk can either be vertical or horizontal.
2. Student diagrams should contain the four corner sidewalks and either the horizontal sidewalk or the vertical sidewalk shown in this diagram.

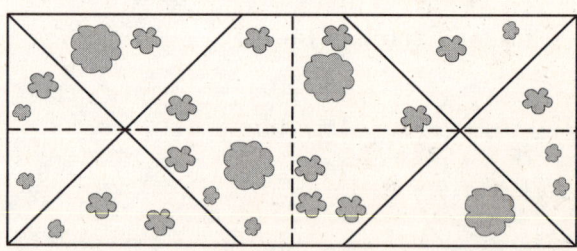

3. Answers depend upon whether students drew a vertical or horizontal sidewalk in Question 1. If they drew a vertical sidewalk, the answer is that the sidewalks drawn in Question 2 intersect in pairs, but do not intersect the sidewalk drawn in Question 1. If they drew a horizontal sidewalk, the answer is that the sidewalks drawn in Question 2 intersect in pairs and the sidewalk drawn in Question 1 intersects all the sidewalks drawn in Question 2. 4. Answers will vary.

Technology Activity
1. 180° 2. 60°

Practice A
1. Check drawings. 2. Check drawings.
3. Check drawings. 4. (1, 2) 5. (1, 1)
6. $\left(-1, -\tfrac{3}{2}\right)$ 7. (2, 3) 8. (1, −2) 9. (2, −2)
10. (4, −4.5) 11. Check drawings.
12. Check drawings. 13. Check drawings.
14. m∠RPT = 19°, m∠RPS = 38°

Lesson 1.5 continued

15. $m\angle TPS = 39°$, $m\angle RPS = 78°$
16. $m\angle RPT = 49°$, $m\angle TPS = 49°$

Practice B
1. Check drawings. 2. Check drawings.
3. $(1, 2)$ 4. $(1, 0)$ 5. $(1, -2.5)$ 6. $(-2, -2)$
7. $(2, -5)$ 8. $(-3, -1)$ 9. Check drawings.
10. Check drawings. 11. Check drawings.
12. $m\angle RPT = 37°$, $m\angle RPS = 74°$
13. $m\angle TPS = 44°$, $m\angle RPS = 88°$
14. $m\angle RPT = 21°$, $m\angle TPS = 21°$
15. 10 16. 5 17. 8

Practice C
1. Check drawings. 2. Check drawings.
3. $(-2, 1)$ 4. $(0.6, 6.4)$ 5. $(1.9, -1.7)$
6. $(2, -1)$ 7. $(1.2, -5)$ 8. $(-3.8, -5.9)$
9. Check drawings. 10. Check drawings.
11. Check drawings.
12. $m\angle TPS = 37°$, $m\angle RPS = 74°$
13. $m\angle RPT = 48°$, $m\angle RPS = 96°$
14. $m\angle RPT = 37.5°$, $m\angle TPS = 37.5°$
15. 15.5 16. $8\frac{1}{3}$ 17. 28

Reteaching with Practice
1. $(2.5, 1.5)$ 2. $(-0.5, 7.5)$ 3. $(-2.5, 1.5)$
4. $(1, -3)$ 5. $(0, 24)$ 6. $(7, 3)$ 7. 5 8. 10

Real-Life Application
1–2.
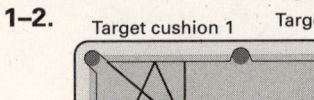
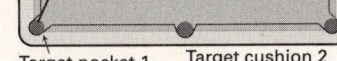

3.
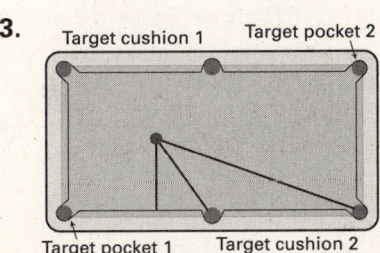

4. It is not possible to make the bank shot because the bisector goes to the side pocket instead of the target cushion.

Challenge: Skills and Applications
1. 5 2. 4 3. yes; *Sample answer:* By the Segment Addition Postulate, $AB + BC = AC$ and $CD + DE = CE$. So, $AB = AC - BC$ and $DE = CE - CD$. But, $AC = CE$ and $BC = CD$ (by the definition of midpoint), so $AB = DE$.
4. $-1, 3$
5. no; *Sample answer:* 6. $22.5°$

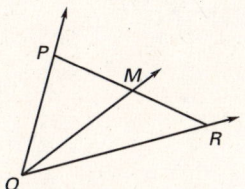

7. $50°, 70°$; 8. $144°$;
Sample answer: *Sample answer:*

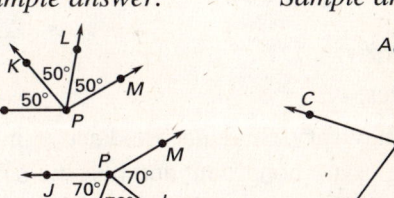

9. $(4, 7, 3)$ 10. $(2, 3, 3)$ 11. $(-7, 4, 5)$
12. $(5, -2.5, 9.5)$ 13. $(4.8, -1.4, 2.1)$
14. $(6.8, 16.3, 5.3)$

Quiz 2
1. If B is the interior of $\angle AOC$, then $m\angle AOB + m\angle BOC = m\angle AOC$.
2–5. coordinates of sample points are given.

2. 3.
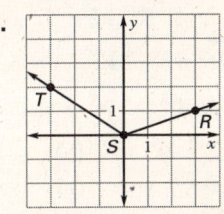

obtuse; $(0, 2), (0, -1)$ right; $(1, 1), (-1, -1)$

Lesson 1.5 continued

4.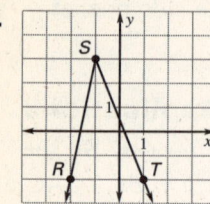
5.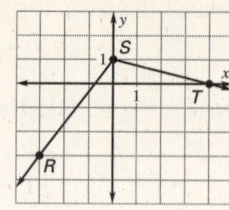

acute; $(-1, -1), (1, 1)$ obtuse; $(0, 0), (2, 2)$

6. $\left(0, \frac{1}{2}\right)$ 7. $15°, 30°$

Lesson 1.6

Warm-Up Exercises
Sample answers are given.
1. $\angle ABC$ 2. $\angle ABE$ 3. $\angle ABD$
4. $\angle ABC$ and $\angle CBD$

Daily Homework Quiz
1. $(-3, 9)$ 2. $(-2, 0)$ 3. $(7, 5)$ 4. 7

Lesson Opener
Allow 10 minutes.

1–3. Answers may vary. Students are discovering that vertical angles are congruent and that the sum of the measures of adjacent angles formed by intersecting lines is 180°.

Practice A
1. yes   2. yes   3. no   4. yes   5. no   6. no
7. 102°   8. 94°   9. 56°   10. 47°   11. 22°
12. 165°   13. 48°, 132°   14. 12°, 102°
15. 73°, 107°   16. 45°, 135°   17. 35   18. 12
19. 16   20. 55   21. 43   22. 31

Practice B
1. yes   2. no   3. no   4. yes   5. 129°
6. 103°   7. 44°   8. 53°   9. 42°, 138°
10. 7°, 97°   11. 53°, 37°   12. 65°, 155°
13. $x = 35, y = 50$ 14. $x = 12, y = 168$
15. $x = 16, y = 10$ 16. $x = 55, y = 105$
17. $x = 43, y = 60$ 18. $x = 31, y = 11$
19. $m\angle A = 70°, m\angle B = 110°, m\angle C = 20°$
20. $m\angle A = 60°, m\angle B = 120°, m\angle C = 30°$

Practice C
1. sometimes   2. always   3. never
4. sometimes   5. $m\angle 7 = 142°, m\angle 8 = 38°$, $m\angle 9 = 142°$ 6. $m\angle 6 = 84°, m\angle 7 = 96°$, $m\angle 9 = 96°$ 7. $m\angle 6 = 44°, m\angle 7 = 136°$, $m\angle 8 = 44°$ 8. $m\angle 6 = 153°, m\angle 8 = 153°$, $m\angle 9 = 27°$ 9. $38°, 142°$ 10. $23°, 113°$
11. $73°, 17°$ 12. $78°, 168°$ 13. $x = 35, y = 12$
14. $x = 35, y = 145$ 15. $x = 10, y = 20$
16. $x = 55, y = 20$ 17. $x = 8, y = 21$
18. $x = 32, y = 29$
19. $m\angle A = 23°, m\angle B = 157°, m\angle C = 67°$
20. $m\angle A = 37°, m\angle B = 143°, m\angle C = 53°$

Reteaching with Practice
1. a. no   b. no   c. yes   d. no   2. a. yes
b. no   c. yes   d. no   3. $x = 15$, $m\angle AEB = 90°, m\angle BEC = 90°, m\angle CED = 90°$, $m\angle AED = 90°$ 4. $y = 40, m\angle FJG = 150°$, $m\angle GJH = 30°, m\angle HJI = 150°, m\angle FJI = 30°$
5. 9°   6. 153°

Interdisciplinary Application
1. $m\angle BDC = 50°$ 2. $m\angle FDE = 50°$
3. $m\angle FDK \approx 35°$ 4. The keys are closer to you than you thought.

Challenge: Skills and Applications
1. *Sample answer:* It does not make sense to say "a vertical angle." Vertical angles are defined only in the context of a *pair* of vertical angles.

2. false; *Sample answer:*

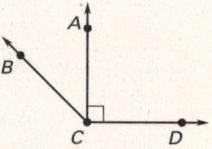

3. true; *Sample answer:* Since $\angle ACD$ is a straight angle, $\overrightarrow{CA}$ and $\overrightarrow{CD}$ are opposite rays. Hence, to show that $\angle ACB$ and $\angle BCD$ are a linear pair (and therefore supplementary), we need only show that $\angle ACB$ and $\angle BCD$ are adjacent angles. Since $\angle ACB$ is an angle, B is not on $\overrightarrow{CA}$; since $\angle BCD$ is an angle, B is not on $\overrightarrow{CD}$. Therefore, B is not on line $\overleftrightarrow{AD}$, so $\angle ACB$ and $\angle BCD$ are two adjacent angles formed by the intersection of lines $\overleftrightarrow{AD}$ and $\overleftrightarrow{BC}$.

Lesson 1.6 continued

4. true; *Sample answer:* Since ∠RUS and ∠SUT are a linear pair, $\overrightarrow{UR}$ and $\overrightarrow{UT}$ are opposite rays. Since ∠SUT and ∠TUV are a linear pair, $\overrightarrow{US}$ and $\overrightarrow{UV}$ are a pair of opposite rays. Therefore, the sides of ∠RUS and ∠TUV form two pairs of opposite rays, so ∠RUS and ∠TUV are vertical angles.

5. false; *Sample answer:*

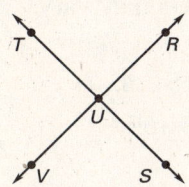

6. $x = 90, y = -40$
7. $x = 8, y = 12$
8. $x = -10, y = 15$
9. $x = 18, y = 8$
10. $x = 13, y = 8$
11. $x = 35, y = -10$

Lesson 1.7

Warm-Up Exercises
1. 6 **2.** $2\sqrt{10}$ **3.** 4 **4.** $2\sqrt{2}$

Daily Homework Quiz
1. no **2.** yes **3.** 60° **4.** 12
5. $m\angle A = 57°; m\angle B = 33°$

Lesson Opener
Allow 15 minutes.

1–3. Estimates for C and A will vary. The answers given were calculated using 3.14 for π.

Circle	d	r	r^2	C	A
A	6	3	9	18.84	28.26
B	4	2	4	12.56	12.56
C	8	4	16	25.12	50.24

4. *Sample answer:* The circumference is a little more than 3 times the diameter. The area is a little more than 3 times the radius squared.

Technology Activity
1. 2 units
2. $2\pi r = \pi r^2$ Let circumference of a circle equal its area.
$2r = r^2$ Divide both sides by π.
$0 = r^2 - 2r$ Subtract $2r$ from both sides.
$0 = r(r - 2)$ Factor.
$0 = r - 2$ Solve for r. ($r = 0$ can be disregarded because it doesn't make sense in the context of the problem.)
$2 = r$

3. a. If the radius is less than 2, then the value of the circumference of a circle is greater than the value of the area of the circle. **b.** If the radius is equal to 2, then the value of the circumference of a circle is equal to the value of the area of the circle. **c.** If the radius is greater than 2, then the value of the circumference of a circle is less than the value of the area of the circle.

Practice A
1. 32 units, 55 square units
2. 16 units, 16 square units
3. 18 units, 12 square units
4. 18.84 units, 28.26 square units
5. 28 units, 49 square units
6. 30 units, 30 square units
7. 54 units, 126 square units
8. 15 units, 12.5 square units
9. 31.4 units, 78.5 square units **10.** 36 cm²
11. 40 in.² **12.** 113.04 ft² **13.** 16 square units
14. 12.56 square units **15.** 9 square units

Practice B
1. 40 units, 91 square units
2. 32 units, 64 square units
3. 36 units, 48 square units
4. 31.4 units, 78.5 square units
5. 48 units, 144 square units
6. 60 units, 120 square units
7. 27 units, 31.5 square units
8. 43.96 units, 153.86 square units
9. 17.2 units, 14.08 square units **10.** 61.5 cm²

Lesson 1.7 continued

11. 50.4 in.² **12.** 1256 ft² **13.** 36 square units
14. 28.26 square units **15.** 12 square units

Practice C
1. $6\frac{2}{3}$ units, $\frac{25}{9}$ square units
2. 23.6 units, 26.97 square units
3. 15.7 units, 19.625 square units
4. 40 units, 60 square units
5. 54 units, 126 square units
6. $8\sqrt{2}$ units, 8 square units
7. $22 + 2\sqrt{37}$ units, $28\sqrt{3}$ square units
8. 33.55 units, 58.875 square units
9. 56 units, 192 square units **10.** 26.66 cm²
11. 67.5 in.² **12.** 1962.5 ft²
13. 28 units, 48 square units
14. 25.12 units, 50.24 square units
15. $7 + 5\sqrt{2} + \sqrt{29}$ units, 17.5 square units
16. 156 bundles **17.** 15386 ft²

Reteaching with Practice
1. 75.36 units, 452.16 square units **2.** 48 units, 84 square units **3.** 21.71 ft, 24.53 ft²

Cooperative Learning Activity
1. Answers may vary. **2.** 30 cm²
3. a, 14 cm; a, 10.4 cm² **4.** 3.42π cm²

Analyzing the Results
1. Answers may vary. **2.** Answers may vary.

Real-Life Application
1. no **2.** slow **3.** 1,263,240 ft²
4. about $13.41 per square foot

Challenge: Skills and Applications
1. 28 units; 36 square units **2.** 73.81 units; 284.955 square units **3.** 69.68 units; 202.08 square units **4.** 24 units; 25 square units
5. 30 units; 22 square units **6.** 16 units; 16 square units **7.** 22.85 units; about 34.81 square units **8.** 28 m; 29 m²

Review and Assessement

Chapter Review Games and Activities
1. point **2.** unproven **3.** midpoint **4.** plane
5. kite **6.** intersect **7.** noncollinear
8. Pythagorean **9.** initial   PUMPKIN PI

Test A
1. 14, 17, 20 **2.** 11, 16, 22
3. 3, 1, $\frac{1}{3}$ **4.** A, E, B **5.** Answers may vary. *Sample answer:* A, E, C **6.** Answers may vary. *Sample answer:* D, E, B, C
7. *Sample answer:* $\overleftrightarrow{DB}$, $\overleftrightarrow{AB}$ **8.** 5 **9.** 7
10. 90° **11.** 110° **12.** 180° **13.** right
14. obtuse **15.** straight
16. (0, 5); 10 **17.** (0, 0); $\sqrt{340}$, or $2\sqrt{85}$
18. (−1, −1); $\sqrt{296}$, or $2\sqrt{74}$
19. $m\angle TEF = 35°$; $m\angle FEA = 35°$
20. $m\angle TEF = 30°$; $m\angle TEA = 60°$
21. 15 **22.** 10 **23.** 25° **24.** 85° **25.** 130°
26. $x = 19$; $y = 69$ **27.** $x = 23$; $y = 13.5$
28. 10 cm² **29.** 9 ft² **30.** 78.5 yd²

Test B
1. 14, 18, 22 **2.** 25, 40, 58
3. $\frac{1}{16}$, $-\frac{1}{32}$, $\frac{1}{64}$ **4.** A, D, B; C, D, E
5. Answers may vary. *Sample answer:* H, A, D
6. Answers may vary. *Sample answer:* H, A, D, C
7. *Sample answer:* $\overleftrightarrow{CE}$ and $\overleftrightarrow{AB}$ **8.** $\sqrt{13}$
9. $7\sqrt{2}$ **10.** 120° **11.** 75° **12.** 60°
13. obtuse **14.** acute
15. straight **16.** (0, −6) **17.** (0, 4) **18.** (3, −9)
19. $m\angle TEF = 37°$, $m\angle AEF = 37°$
20. $m\angle TEF = 29°$, $m\angle TEA = 58°$
21. 36 **22.** 4 **23.** 44° **24.** 102° **25.** 33°
26. vertical angles **27.** linear pair
28. vertical angles **29.** $x = 25$; $y = 29$
30. $x = 23$; $y = 60$ **31.** 10 cm² **32.** 49 ft²
33. 50.24 yd² (using 3.14)

Review and Assessment *continued*

Test C
1. 50, 250, 1250 2. 8, 13, 19
3. −52, −77, −107 4. Answers may vary. Sample answer: F, A, B 5. E, B, D
6. Answers may vary. Sample answer: A, B, F, E
7. Sample answer: $\overleftrightarrow{AB}, \overleftrightarrow{BD}$ 8. $\sqrt{26}$
9. $\sqrt{41}$ 10. 140° 11. 22° 12. 90°
13. obtuse 14. acute 15. right
16. (2, −2) 17. (1, 2) 18. (−2, −6)
19. $m\angle TEA = 74.5°; m\angle TEF = 37.25°$
20. $m\angle TEF = 43.5°; m\angle FEA = 43.5°$
21. 15 22. 17 23. 37° 24. 65°
25. $(180 − 2x)°$ 26. vertical angles
27. neither 28. linear pair 29. $x = 43; y = 77$
30. $x = 17.5; y = 28.5$ 31. 8.125 cm²
32. 20.25 ft² 33. 113.04 yd² (using 3.14) or 113.10 yd² (using π)

SAT/ACT Test
1. B 2. D 3. C 4. A 5. B 6. A 7. C
8. C 9. D

Alternative Assessment
1. Complete answers should include:
- A diagram with E, F, and G not collinear.
- Also, $\overleftrightarrow{AE}$ intersecting $\overleftrightarrow{BG}$ at the point F.
- Two opposite rays: $\overrightarrow{FB}$ and $\overrightarrow{FG}$ or $\overrightarrow{FE}$ and $\overrightarrow{FA}$.
- Vertical angles: $\angle BFA$ and $\angle GFE$ or $\angle BFE$ and $\angle GFA$. • Supplementary angles: $\angle BFA$ and $\angle AFG$, or $\angle GFE$ and $\angle EFB$, or $\angle AFG$ and $\angle GFE$, or $\angle EFB$ and $\angle BFA$
- $m\angle AFB + m\angle EFB = 180°$

2. a. 26 units; M(3, 7) b. 7 units c. 76°
d. 90° e. Answers may vary. Sample answer: $\angle TAN; \angle TAP; \angle TAK; \angle SAP$ f. 45°
3. a. About 27.7 units b. About 57.7 units and 204 square units 4. Answers may vary. Answers should include acute angle, right angle, obtuse angle, and straight angle. Answers should have a diagram and an explanation of each classification.

Project: Creating Paper Measuring Tools

1. A sample labeling of points is shown in the diagram. The explanations in Exercises 2 and 3 use that labeling in explaining how to find the angle measures. Accept less specific names for the shapes in the seven regions.

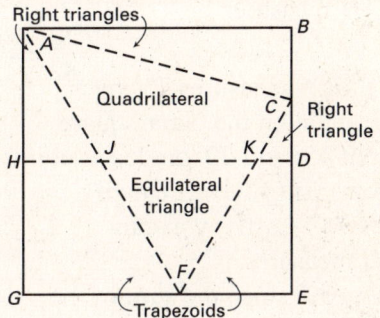

2.–3.

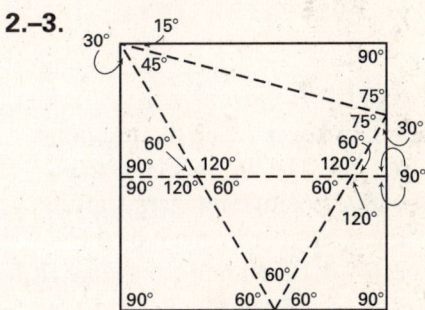

See below for explanations.

2. *Sample reasoning*:
- $\angle B, \angle E,$ and $\angle G$ each measure 90° since *ABEG* is a square.
- The two angles at point *D* and the two angles at point *H* measure 90° since the horizontal fold bisects the straight angles at those points.
- By the construction, the three angles at point *F* are congruent. The sum of the measures of these three angles is 180° so the measure of each angle is 60°.
- $m\angle BAC + m\angle CAJ + m\angle JAH = 90°$. By folding, $\angle CAJ \cong \angle BAC + \angle JAH$. Therefore, by substitution of equals, $m\angle CAJ + m\angle CAJ = 90°$ so $\angle CAJ$ measures 45°.

3. *Sample reasoning*: • Using a 60° angle from the other square, you can see that one of the acute angles at point *J* measures 60°. The two acute angles form a vertical pair so the other acute angle also measures 60°. The acute and obtuse angles are supplementary so each obtuse angle measures $180° − 60° = 120°$.

Review and Assessment continued

- By repeating the steps above, you find that there are two 60° angles and two 120° at point K.
- Using a 60° angle from the other square, you can see that ∠BAJ measures 60°. ∠CAJ measures 45° (see answer to Ex. 2), so ∠BAC measures 60° − 45° = 15°. Since ∠BAJ and ∠JAH are complementary, ∠JAH measures 90° − 60° = 30°.
- You can find that ∠BCA measures 75° by placing against it the angle on the other square formed by the adjacent 45° and 30° angles at point A. By the construction, ∠KCA ≅ ∠BCA so ∠KCA also measures 75°. Lastly, you can conclude that ∠KCD measures 180° − (75° + 75°) = 30° since the sum of the measures of the three angles at point E must be 180°.

Cumulative Review

1. *Sample:* Each number is twice the previous number plus 2; 62   **2.** The numbers are the squares of consecutive positive integers starting with 1; 25   **3.** Each number is one half the previous number; 6   **4.** Each number is one fifth the previous number; 5   **5.** even   **6.** odd
7. false   **8.** true   **9.** false   **10.** true   **11.** true
12. true   **13.** $AB = 10$, $BC = 15$, $CD = 5$
14. 5.39   **15.** 4.47   **16.** 6.08   **17.** 7.07
18. 9.22   **19.** 6.40
20.   **21.**

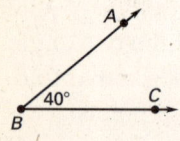

22. 70°   **23.** 135°   **24.** 70°   **25.** 40°
26. 160°   **27.** 130°   **28.** (8, 1)   **29.** (−1, 8)
30. $\left(\frac{5}{2}, -3\right)$ **31.** $\left(\frac{15}{2}, 1\right)$
32. $m\angle A = 57°$, $m\angle B = 33°$
33. $m\angle A = 71°$, $m\angle B = 19°$ **34.** 125.6 ft
35. 18 yd   **36.** 48 ft²   **37.** 12 yd²

CURRICULUM
MSSU